50% OFF Online TSI Prep Course

Dear Customer,

We consider it an honor and a privilege that you chose our TSI Study Guide. As a way of showing our appreciation and to help us better serve you, we have partnered with Mometrix Test Preparation to offer **50% off their online TSI Prep Course**. Many TSI courses are needlessly expensive and don't deliver enough value. With our course, you get access to the best TSI prep material, and you only pay half price.

Mometrix has structured their online course to perfectly complement your printed study guide. The TSI Prep Course contains **over 70 lessons** that cover all the most important topics, **110+ video reviews** that explain difficult concepts, **over 600 practice questions** to ensure you feel prepared, and **digital flashcards**, so you can study while you're on the go.

Online TSI Prep Course

Topics Covered:

- Mathematics
  - *Elementary Algebra and Functions*
  - *Intermediate Algebra and Functions*
  - *Geometry and Measurement*
  - *Data Analysis, Statistics, and Probability*
- Reading
  - *Literary Analysis*
  - *Main Idea and Supporting Details*
  - *Inferences in a Text or Texts*
  - *Author's Use of Language*
- Writing
  - *Foundations of Grammar and Punctuation*
  - *Essay Revision and Sentence Logic*
  - *Agreement and Sentence Structure*

Course Features:

- TSI Study Guide
  - Get content that complements our best-selling study guide.
- 9 Full-Length Practice Tests
  - With over 600 practice questions, you can test yourself again and again.
- Mobile Friendly
  - If you need to study on the go, the course is easily accessible from your mobile device.
- TSI Flashcards
  - Our course includes a flashcard mode consisting of over 600 content cards to help you study.

To receive this discount, simply head to their website: www.mometrix.com/university/courses/tsi and add the course to your cart. At the checkout page, enter the discount code: **TPBTSI50**

If you have any questions or concerns, please don't hesitate to contact Mometrix at universityhelp@mometrix.com.

Sincerely,

FREE Test Taking Tips DVD Offer

To help us better serve you, we have developed a Test Taking Tips DVD that we would like to give you for FREE. **This DVD covers world-class test taking tips that you can use to be even more successful when you are taking your test.**

All that we ask is that you email us your feedback about your study guide. Please let us know what you thought about it – whether that is good, bad or indifferent.

To get your **FREE Test Taking Tips DVD**, email freedvd@studyguideteam.com with "FREE DVD" in the subject line and the following information in the body of the email:

      a. The title of your study guide.

      b. Your product rating on a scale of 1-5, with 5 being the highest rating.

      c. Your feedback about the study guide. What did you think of it?

      d. Your full name and shipping address to send your free DVD.

If you have any questions or concerns, please don't hesitate to contact us at freedvd@studyguideteam.com.

Thanks again!

TSI Study Guide 2020-2021

TSI Prep and 3 Complete Practice Tests for the Texas Success Initiative [4th Edition]

TPB Publishing

Interested in buying more than 10 copies of our product? Contact us about bulk discounts: bulkorders@studyguideteam.com

ISBN 13: 9781628456998
ISBN 10: 162845699X

Table of Contents

//Test Prep Books|||

Quick Overview

As you draw closer to taking your exam, effective preparation becomes more and more important. Thankfully, you have this study guide to help you get ready. Use this guide to help keep your studying on track and refer to it often.

This study guide contains several key sections that will help you be successful on your exam. The guide contains tips for what you should do the night before and the day of the test. Also included are test-taking tips. Knowing the right information is not always enough. Many well-prepared test takers struggle with exams. These tips will help equip you to accurately read, assess, and answer test questions.

A large part of the guide is devoted to showing you what content to expect on the exam and to helping you better understand that content. In this guide are practice test questions so that you can see how well you have grasped the content. Then, answer explanations are provided so that you can understand why you missed certain questions.

Don't try to cram the night before you take your exam. This is not a wise strategy for a few reasons. First, your retention of the information will be low. Your time would be better used by reviewing information you already know rather than trying to learn a lot of new information. Second, you will likely become stressed as you try to gain a large amount of knowledge in a short amount of time. Third, you will be depriving yourself of sleep. So be sure to go to bed at a reasonable time the night before. Being well-rested helps you focus and remain calm.

Be sure to eat a substantial breakfast the morning of the exam. If you are taking the exam in the afternoon, be sure to have a good lunch as well. Being hungry is distracting and can make it difficult to focus. You have hopefully spent lots of time preparing for the exam. Don't let an empty stomach get in the way of success!

When travelling to the testing center, leave earlier than needed. That way, you have a buffer in case you experience any delays. This will help you remain calm and will keep you from missing your appointment time at the testing center.

Be sure to pace yourself during the exam. Don't try to rush through the exam. There is no need to risk performing poorly on the exam just so you can leave the testing center early. Allow yourself to use all of the allotted time if needed.

Remain positive while taking the exam even if you feel like you are performing poorly. Thinking about the content you should have mastered will not help you perform better on the exam.

Once the exam is complete, take some time to relax. Even if you feel that you need to take the exam again, you will be well served by some down time before you begin studying again. It's often easier to convince yourself to study if you know that it will come with a reward!

Test-Taking Strategies

1. Predicting the Answer

When you feel confident in your preparation for a multiple-choice test, try predicting the answer before reading the answer choices. This is especially useful on questions that test objective factual knowledge. By predicting the answer before reading the available choices, you eliminate the possibility that you will be distracted or led astray by an incorrect answer choice. You will feel more confident in your selection if you read the question, predict the answer, and then find your prediction among the answer choices. After using this strategy, be sure to still read all of the answer choices carefully and completely. If you feel unprepared, you should not attempt to predict the answers. This would be a waste of time and an opportunity for your mind to wander in the wrong direction.

2. Reading the Whole Question

Too often, test takers scan a multiple-choice question, recognize a few familiar words, and immediately jump to the answer choices. Test authors are aware of this common impatience, and they will sometimes prey upon it. For instance, a test author might subtly turn the question into a negative, or he or she might redirect the focus of the question right at the end. The only way to avoid falling into these traps is to read the entirety of the question carefully before reading the answer choices.

3. Looking for Wrong Answers

Long and complicated multiple-choice questions can be intimidating. One way to simplify a difficult multiple-choice question is to eliminate all of the answer choices that are clearly wrong. In most sets of answers, there will be at least one selection that can be dismissed right away. If the test is administered on paper, the test taker could draw a line through it to indicate that it may be ignored; otherwise, the test taker will have to perform this operation mentally or on scratch paper. In either case, once the obviously incorrect answers have been eliminated, the remaining choices may be considered. Sometimes identifying the clearly wrong answers will give the test taker some information about the correct answer. For instance, if one of the remaining answer choices is a direct opposite of one of the eliminated answer choices, it may well be the correct answer. The opposite of obviously wrong is obviously right! Of course, this is not always the case. Some answers are obviously incorrect simply because they are irrelevant to the question being asked. Still, identifying and eliminating some incorrect answer choices is a good way to simplify a multiple-choice question.

4. Don't Overanalyze

Anxious test takers often overanalyze questions. When you are nervous, your brain will often run wild, causing you to make associations and discover clues that don't actually exist. If you feel that this may be a problem for you, do whatever you can to slow down during the test. Try taking a deep breath or counting to ten. As you read and consider the question, restrict yourself to the particular words used by the author. Avoid thought tangents about what the author *really* meant, or what he or she was *trying* to say. The only things that matter on a multiple-choice test are the words that are actually in the question. You must avoid reading too much into a multiple-choice question, or supposing that the writer meant something other than what he or she wrote.

5. No Need for Panic

It is wise to learn as many strategies as possible before taking a multiple-choice test, but it is likely that you will come across a few questions for which you simply don't know the answer. In this situation, avoid panicking. Because most multiple-choice tests include dozens of questions, the relative value of a single wrong answer is small. As much as possible, you should compartmentalize each question on a multiple-choice test. In other words, you should not allow your feelings about one question to affect your success on the others. When you find a question that you either don't understand or don't know how to answer, just take a deep breath and do your best. Read the entire question slowly and carefully. Try rephrasing the question a couple of different ways. Then, read all of the answer choices carefully. After eliminating obviously wrong answers, make a selection and move on to the next question.

6. Confusing Answer Choices

When working on a difficult multiple-choice question, there may be a tendency to focus on the answer choices that are the easiest to understand. Many people, whether consciously or not, gravitate to the answer choices that require the least concentration, knowledge, and memory. This is a mistake. When you come across an answer choice that is confusing, you should give it extra attention. A question might be confusing because you do not know the subject matter to which it refers. If this is the case, don't eliminate the answer before you have affirmatively settled on another. When you come across an answer choice of this type, set it aside as you look at the remaining choices. If you can confidently assert that one of the other choices is correct, you can leave the confusing answer aside. Otherwise, you will need to take a moment to try to better understand the confusing answer choice. Rephrasing is one way to tease out the sense of a confusing answer choice.

7. Your First Instinct

Many people struggle with multiple-choice tests because they overthink the questions. If you have studied sufficiently for the test, you should be prepared to trust your first instinct once you have carefully and completely read the question and all of the answer choices. There is a great deal of research suggesting that the mind can come to the correct conclusion very quickly once it has obtained all of the relevant information. At times, it may seem to you as if your intuition is working faster even than your reasoning mind. This may in fact be true. The knowledge you obtain while studying may be retrieved from your subconscious before you have a chance to work out the associations that support it. Verify your instinct by working out the reasons that it should be trusted.

8. Key Words

Many test takers struggle with multiple-choice questions because they have poor reading comprehension skills. Quickly reading and understanding a multiple-choice question requires a mixture of skill and experience. To help with this, try jotting down a few key words and phrases on a piece of scrap paper. Doing this concentrates the process of reading and forces the mind to weigh the relative importance of the question's parts. In selecting words and phrases to write down, the test taker thinks about the question more deeply and carefully. This is especially true for multiple-choice questions that are preceded by a long prompt.

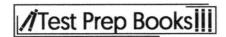

9. Subtle Negatives

One of the oldest tricks in the multiple-choice test writer's book is to subtly reverse the meaning of a question with a word like *not* or *except*. If you are not paying attention to each word in the question, you can easily be led astray by this trick. For instance, a common question format is, "Which of the following is...?" Obviously, if the question instead is, "Which of the following is not...?," then the answer will be quite different. Even worse, the test makers are aware of the potential for this mistake and will include one answer choice that would be correct if the question were not negated or reversed. A test taker who misses the reversal will find what he or she believes to be a correct answer and will be so confident that he or she will fail to reread the question and discover the original error. The only way to avoid this is to practice a wide variety of multiple-choice questions and to pay close attention to each and every word.

10. Reading Every Answer Choice

It may seem obvious, but you should always read every one of the answer choices! Too many test takers fall into the habit of scanning the question and assuming that they understand the question because they recognize a few key words. From there, they pick the first answer choice that answers the question they believe they have read. Test takers who read all of the answer choices might discover that one of the latter answer choices is actually *more* correct. Moreover, reading all of the answer choices can remind you of facts related to the question that can help you arrive at the correct answer. Sometimes, a misstatement or incorrect detail in one of the latter answer choices will trigger your memory of the subject and will enable you to find the right answer. Failing to read all of the answer choices is like not reading all of the items on a restaurant menu: you might miss out on the perfect choice.

11. Spot the Hedges

One of the keys to success on multiple-choice tests is paying close attention to every word. This is never truer than with words like almost, most, some, and sometimes. These words are called "hedges" because they indicate that a statement is not totally true or not true in every place and time. An absolute statement will contain no hedges, but in many subjects, the answers are not always straightforward or absolute. There are always exceptions to the rules in these subjects. For this reason, you should favor those multiple-choice questions that contain hedging language. The presence of qualifying words indicates that the author is taking special care with his or her words, which is certainly important when composing the right answer. After all, there are many ways to be wrong, but there is only one way to be right! For this reason, it is wise to avoid answers that are absolute when taking a multiple-choice test. An absolute answer is one that says things are either all one way or all another. They often include words like *every*, *always*, *best*, and *never*. If you are taking a multiple-choice test in a subject that doesn't lend itself to absolute answers, be on your guard if you see any of these words.

12. Long Answers

In many subject areas, the answers are not simple. As already mentioned, the right answer often requires hedges. Another common feature of the answers to a complex or subjective question are qualifying clauses, which are groups of words that subtly modify the meaning of the sentence. If the question or answer choice describes a rule to which there are exceptions or the subject matter is complicated, ambiguous, or confusing, the correct answer will require many words in order to be expressed clearly and accurately. In essence, you should not be deterred by answer choices that seem excessively long. Oftentimes, the author of the text will not be able to write the correct answer without

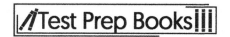

offering some qualifications and modifications. Your job is to read the answer choices thoroughly and completely and to select the one that most accurately and precisely answers the question.

13. Restating to Understand

Sometimes, a question on a multiple-choice test is difficult not because of what it asks but because of how it is written. If this is the case, restate the question or answer choice in different words. This process serves a couple of important purposes. First, it forces you to concentrate on the core of the question. In order to rephrase the question accurately, you have to understand it well. Rephrasing the question will concentrate your mind on the key words and ideas. Second, it will present the information to your mind in a fresh way. This process may trigger your memory and render some useful scrap of information picked up while studying.

14. True Statements

Sometimes an answer choice will be true in itself, but it does not answer the question. This is one of the main reasons why it is essential to read the question carefully and completely before proceeding to the answer choices. Too often, test takers skip ahead to the answer choices and look for true statements. Having found one of these, they are content to select it without reference to the question above. Obviously, this provides an easy way for test makers to play tricks. The savvy test taker will always read the entire question before turning to the answer choices. Then, having settled on a correct answer choice, he or she will refer to the original question and ensure that the selected answer is relevant. The mistake of choosing a correct-but-irrelevant answer choice is especially common on questions related to specific pieces of objective knowledge. A prepared test taker will have a wealth of factual knowledge at his or her disposal, and should not be careless in its application.

15. No Patterns

One of the more dangerous ideas that circulates about multiple-choice tests is that the correct answers tend to fall into patterns. These erroneous ideas range from a belief that B and C are the most common right answers, to the idea that an unprepared test-taker should answer "A-B-A-C-A-D-A-B-A." It cannot be emphasized enough that pattern-seeking of this type is exactly the WRONG way to approach a multiple-choice test. To begin with, it is highly unlikely that the test maker will plot the correct answers according to some predetermined pattern. The questions are scrambled and delivered in a random order. Furthermore, even if the test maker was following a pattern in the assignation of correct answers, there is no reason why the test taker would know which pattern he or she was using. Any attempt to discern a pattern in the answer choices is a waste of time and a distraction from the real work of taking the test. A test taker would be much better served by extra preparation before the test than by reliance on a pattern in the answers.

FREE DVD OFFER

Don't forget that doing well on your exam includes both understanding the test content and understanding how to use what you know to do well on the test. We offer a completely FREE Test Taking Tips DVD that covers world class test taking tips that you can use to be even more successful when you are taking your test.

All that we ask is that you email us your feedback about your study guide. To get your **FREE Test Taking Tips DVD**, email freedvd@studyguideteam.com with "FREE DVD" in the subject line and the following information in the body of the email:

- The title of your study guide.
- Your product rating on a scale of 1-5, with 5 being the highest rating.
- Your feedback about the study guide. What did you think of it?
- Your full name and shipping address to send your free DVD.

Introduction to the TSI Exam

Function of the Test

The Texas Success Initiative Assessment (TSI) is used to measure a student's readiness for college-level coursework. The TSI assesses incoming freshman students in the state of Texas in the subject areas of reading, writing and mathematics. The results of this test are used to place an incoming student into the appropriate college-level course and determine the type of intervention, if any, a student needs.

Many students can be exempt from the TSI if they met the minimum score on the SAT, ACT or statewide high school test. If students have already completed college-level English and math courses, they will be exempt. Other ways to be exempt from taking the TSI include current enrollment in a Level-One certificate program (fewer than 43 semester credit hours), not seeking a college degree, or students currently or formerly enrolled in the military. Those students that are not exempt will be asked to take three tests, one in each of the subject areas of mathematics, reading, and writing. A diagnostic test may be given in addition to these three tests in a particular subject area if deemed necessary.

The TSI is meant to give an institute detailed information on a student's strengths and weaknesses. Before taking the TSI, a candidate must take the Pre-Assessment activity. The activity includes an explanation of the TSI, practice questions and feedback, an explanation of developmental educational options if a minimum score is not met, and campus and community resources.

Test Administration

There are testing centers located at universities for the TSI assessment. Retesting is acceptable, and there are no restrictions as to when a retest can be done. However, students are encouraged to study and prepare before retaking the TSI. Some colleges and universities also host workshops to better prepare a student to retake the TSI. It is advised to contact the local university or college where a candidate wishes to take the test to receive more information on retesting and workshops. Advisors and counselors at a particular institution can better help a candidate with any further questions they may have.

Test Format

The TSI has three different components: one in mathematics, one in reading and one in writing. All sections are multiple choice with an additional essay section for writing. The table below outlines the concepts tested in each of the sections.

Mathematics	Reading	Writing
Elementary Algebra and Functions	Literary Analysis	Essay Revision
Intermediate Algebra and Functions	Main Idea and Supporting Details	Agreement
Geometry and Measurement	Inferences in a Text or Texts	Sentence Structure
Data Analysis, Statistics, and Probability	Author's Use of Language	Sentence Logic

The essay section is a five paragraph persuasive essay on a controversial issue or current event topic. The essay should be 300-600 words and must clearly state a main idea with specific evidence to support the main idea. Correct conventions are expected, and no dictionaries or resources may be used. Scratch paper and essay paper will be provided at the testing center. Students are not allowed to use their own calculators, but a pop-up calculator will be available on select questions. Students are not allowed to use their own calculators, but a pop-up calculator will be available on select questions.

Section	Approximate Number of Items on Placement Test*	Approximate Number of Items on Diagnostic Test
Mathematics	20	10
Reading	24	10-12
Writing	20	10-12

*Note that the Placement Test is the standard test that is taken by most test candidates. The Diagnostic Test is usually required for those candidates who fail to meet the college-readiness cut score in any of the three domains after taking the Placement Test. Results from the Diagnostic Test can then be used to help place the student in an appropriate developmental or readiness course.

Scoring

The minimum passing score for the TSI Assessment to determine a student's readiness for freshman college-level coursework is different for each subject area. A TSI score of 351 or above is necessary for reading. A TSI score of at least 350 and an essay score of at least 5 is required for writing. Other ways to meet requirements for writing would be a score of at least 363 and an essay score of 4 or a score with less than 350 but with an ABE diagnostic level of at least 4 and an essay score of at least 5. A TSI score of 350 or above is required for math.

Mathematics

Elementary Algebra and Functions

Relations and Functions

First, it's important to understand the definition of a *relation*. Given two variables, x and y, which stand for unknown numbers, a *relation* between x and y is an object that splits all of the pairs (x, y) into those for which the relation is true and those for which it is false. For example, consider the relation of $x^2 = y^2$. This relationship is true for the pair (1, 1) and for the pair (-2, 2), but false for (2, 3). Another example of a relation is $x \leq y$. This is true whenever x is less than or equal to y.

A *function* is a special kind of relation where, for each value of x, there is only a single value of y that satisfies the relation. So, $x^2 = y^2$ is *not* a function because in this case, if x is 1, y can be either 1 or -1: the pair (1, 1) and (1, -1) both satisfy the relation. More generally, for this relation, any pair of the form $(a, \pm a)$ will satisfy it. On the other hand, consider the following relation: $y = x^2 + 1$. This is a function because for each value of x, there is a unique value of y that satisfies the relation. Notice, however, there are multiple values of x that give us the same value of y. This is perfectly acceptable for a function. Therefore, y is a function of x.

To determine if a relation is a function, check to see if every x value has a unique corresponding y value.

A function can be viewed as an object that has x as its input and outputs a unique y-value. It is sometimes convenient to express this using *function notation*, where the function itself is given a name, often f. To emphasize that f takes x as its input, the function is written as $f(x)$. In the above example, the equation could be rewritten as $f(x) = x^2 + 1$. To write the value that a function yields for some specific value of x, that value is put in place of x in the function notation. For example, $f(3)$ means the value that the function outputs when the input value is 3. If $f(x) = x^2 + 1$, then $f(3) = 3^2 + 1 = 10$.

A function can also be viewed as a table of pairs (x, y), which lists the value for y for each possible value of x.

The set of all possible values for x in $f(x)$ is called the *domain* of the function, and the set of all possible outputs is called the *range* of the function. Note that usually the domain is assumed to be all real numbers, except those for which the expression for $f(x)$ is not defined, unless the problem specifies otherwise. An example of how a function might not be defined is in the case of $f(x) = \frac{1}{x+1}$, which is not defined when $x = -1$ (which would require dividing by zero). Therefore, in this case the domain would be all real numbers except $x = -1$.

If y is a function of x, then x is the *independent variable* and y is the *dependent variable*. This is because in many cases, the problem will start with some value of x and then see how y changes depending on this starting value.

Evaluating Functions

To evaluate functions, plug in the given value everywhere the variable appears in the expression for the function. For example, find $g(-2)$ where $g(x) = 2x^2 - \frac{4}{x}$. To complete the problem, plug in -2 in the following way: $g(-2) = 2(-2)^2 - \frac{4}{-2} = 2 \times 4 + 2 = 8 + 2 = 10$.

Defining Linear Equations

A function is called *linear* if it can take the form of the equation $f(x) = ax + b$, or $y = ax + b$, for any two numbers a and b. A linear equation forms a straight line when graphed on the coordinate plane. An example of a linear function is shown below on the graph.

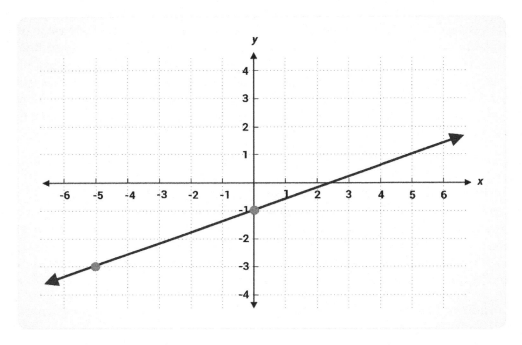

This is a graph of the following function: $y = \frac{2}{5}x - 1$. A table of values that satisfies this function is shown below.

x	y
-5	-3
0	-1
5	1
10	3

These points can be found on the graph using the form (x,y). For more on graphing in the coordinate plane, refer to the Graphing section below.

Graphing Functions and Relations

To graph relations and functions, the Cartesian plane is used. This means to think of the plane as being given a grid of squares, with one direction being the x-axis and the other direction the y-axis. Generally, the independent variable is placed along the horizontal axis, and the dependent variable is placed along the vertical axis. Any point on the plane can be specified by saying how far to go along the x-axis and how far along the y-axis with a pair of numbers (x, y). Specific values for these pairs can be given names such as $C = (-1, 3)$. Negative values mean to move left or down; positive values mean to move right or up. The point where the axes cross one another is called the *origin*. The origin has coordinates $(0, 0)$ and is usually called O when given a specific label. An illustration of the Cartesian plane, along with graphs of $(2, 1)$ and $(-1, -1)$, are below.

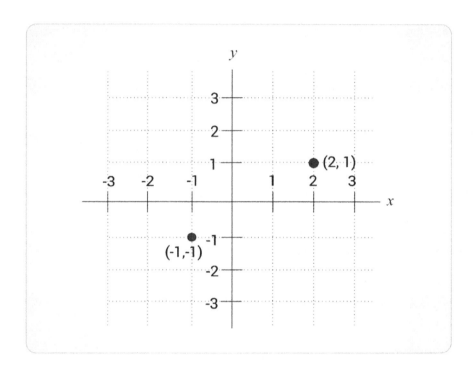

Relations also can be graphed by marking each point whose coordinates satisfy the relation. If the relation is a function, then there is only one value of y for any given value of x. This leads to the **vertical line test**: if a relation is graphed, then the relation is a function if any possible vertical line drawn anywhere along the graph would only touch the graph of the relation in no more than one place. Conversely, when graphing a function, then any possible vertical line drawn will not touch the graph of the function at any point or will touch the function at just one point. This test is made from the definition of a function, where each x-value must be mapped to one and only one y-value.

Forms of Linear Equations

When graphing a linear function, note that the ratio of the change of the y coordinate to the change in the x coordinate is constant between any two points on the resulting line, no matter which two points are chosen. In other words, in a pair of points on a line, (x_1, y_1) and (x_2, y_2), with $x_1 \neq x_2$ so that the two points are distinct, then the ratio $\frac{y_2-y_1}{x_2-x_1}$ will be the same, regardless of which particular pair of points are chosen. This ratio, $\frac{y_2-y_1}{x_2-x_1}$, is called the *slope* of the line and is frequently denoted with the letter m. If slope m is positive, then the line goes upward when moving to the right, while if slope m is negative, then the line goes downward when moving to the right. If the slope is 0, then the line is called *horizontal*, and the y coordinate is constant along the entire line. In lines where the x coordinate is constant along the entire line, y is not actually a function of x. For such lines, the slope is not defined. These lines are called *vertical* lines.

Linear functions may take forms other than $y = ax + b$. The most common forms of linear equations are explained below:

1. Standard Form: $Ax + By = C$, in which the slope is given by $m = \frac{-A}{B}$, and the y-intercept is given by $\frac{C}{B}$.

2. Slope-Intercept Form: $y = mx + b$, where the slope is m and the y intercept is b.

3. Point-Slope Form: $y - y_1 = m(x - x_1)$, where the slope is m and (x_1, y_1) is any point on the chosen line.

4. Two-Point Form: $\frac{y-y_1}{x-x_1} = \frac{y_2-y_1}{x_2-x_1}$, where (x_1, y_1) and (x_2, y_2) are any two distinct points on the chosen line. Note that the slope is given by $m = \frac{y_2-y_1}{x_2-x_1}$.

5. Intercept Form: $\frac{x}{x_1} + \frac{y}{y_1} = 1$, in which x_1 is the x-intercept and y_1 is the y-intercept.

These five ways to write linear equations are all useful in different circumstances. Depending on the given information, it may be easier to write one of the forms over another.

If $y = mx$, y is directly proportional to x. In this case, changing x by a factor changes y by that same factor. If $y = \frac{m}{x}$, y is inversely proportional to x. For example, if x is increased by a factor of 3, then y will be decreased by the same factor, 3.

Solving Linear Equations

Sometimes, rather than a situation where there's an equation such as $y = ax + b$ and finding y for some value of x is requested, the result is given and finding x is requested.

The key to solving any equation is to remember that from one true equation, another true equation can be found by adding, subtracting, multiplying, or dividing both sides by the same quantity. In this case, it's necessary to manipulate the equation so that one side only contains x. Then the other side will show what x is equal to.

For example, in solving $3x - 5 = 2$, adding 5 to each side results in $3x = 7$. Next, dividing both sides by 3 results in $x = \frac{7}{3}$. To ensure the value of x is correct, the number can be substituted into the original equation and solved to see if it makes a true statement. For example, $3(\frac{7}{3}) - 5 = 2$ can be simplified by cancelling out the two 3s. This yields $7 - 5 = 2$, which is a true statement.

Sometimes an equation may have more than one x term. For example, consider the following equation: $3x + 2 = x - 4$. Moving all of the x terms to one side by subtracting x from both sides results in $2x + 2 = -4$. Next, subtract 2 from both sides so that there is no constant term on the left side. This yields $2x = -6$. Finally, divide both sides by 2, which leaves $x = -3$.

Solving Linear Inequalities

Solving linear inequalities is very similar to solving equations, except for one rule: when multiplying or dividing an inequality by a negative number, the inequality symbol changes direction. Given the following inequality, solve for x: $-2x + 5 < 13$. The first step in solving this equation is to subtract 5 from both sides. This leaves the inequality: $-2x < 8$. The last step is to divide both sides by -2. By using the rule, the answer to the inequality is $x > -4$.

Since solutions to inequalities include more than one value, number lines are used many times to model the answer. For the previous example, the answer is modelled on the number line below. It shows that any number greater than -4, not including -4, satisfies the inequality.

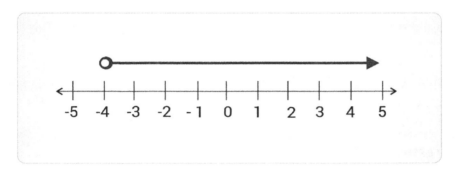

Linear Systems of Equations

A problem sometimes involves multiple variables and multiple equations. These are called *systems of equations*. In this case, try to manipulate them until an expression is found that provides the value of one of the variables. There are a couple of different approaches to this, and some of them can be used together in some cases. The three basic rules to keep in mind are the following.

1. Manipulate a set of equations by doing the same operation to both equations, just as is done when working with just one equation.

2. If one of the equations can be changed so that it expresses one variable in terms of the others, then that expression can be substituted into the other equations and the variable can be eliminated. This means the other equations will have one less variable in them. This is called the method of substitution.

3. If two equations of the form $a = b, c = d$ are included, then a new equation can be formed by adding the left sides and adding the right sides, $a + c = b + d$, or $a - c = b - d$. This enables the elimination of one of the variables from an equation. This is called the method of elimination.

The simplest case is the case of a *linear* system of equations. Although the equations may be written in more complicated forms, linear systems of equations with two variables can always be written in the form $ax + by = c, dx + ey = f$. The two basic approaches to solving these systems are substitution and elimination.

Consider the system $3x - y = 2$ and $2x + 2y = 3$. This can be solved in two ways.

1. By substitution: start by solving the first equation for y. First, subtract $3x$ from both sides to obtain $-y = 2 - 3x$. Next, divide both sides by -1, to obtain $y = 3x - 2$. Then substitute this value for y into the second equation. This yields $2x + 2(3x - 2) = 3$. This can be simplified to $2x + 6x - 4 = 3$, or $8x = 7$, which means $x = \frac{7}{8}$. By plugging in this value for x into $y = 3x - 2$, the result is $y = 3\left(\frac{7}{8}\right) - 2 = \frac{21}{8} - \frac{16}{8} = \frac{5}{8}$. So, this results in $x = \frac{7}{8}, y = \frac{5}{8}$.

2. By elimination: first, multiply the first equation by 2. This results in $-2y$, which could cancel out the $+2y$ in the second equation. Multiplying both sides of the first equation by 2 gives results in $2(3x - y) = 2(2)$, or $6x - 2y = 4$. Adding the left sides and the right sides of the two equations and setting the results equal to one another results in $(6x + 2x) + (-2y + 2y) = 4 + 3$. This simplifies to $8x = 7$, so again $x = \frac{7}{8}$. Plug this back into either of the original equations and the result is $3\left(\frac{7}{8}\right) - y = 2$ or $y = 3\left(\frac{7}{8}\right) - 2 = \frac{21}{8} - \frac{16}{8} = \frac{5}{8}$. This again yields $x = \frac{7}{8}, y = \frac{5}{8}$.

As this shows, both methods will give the same answer. However, one method is sometimes preferred over another simply because of the amount of work required. To check the answer, the values can be substituted into the given system to make sure they form two true statements.

Algebraic Expressions and Equations

Algebraic expressions look similar to equations, but they do not include the equal sign. Algebraic expressions are comprised of numbers, variables, and mathematical operations. Some examples of algebraic expressions are $8x + 7y - 12z$, $3a^2$, and $5x^3 - 4y^4$.

Algebraic expressions and equations can be used to represent real-life situations and model the behavior of different variables. For example, $2x + 5$ could represent the cost to play games at an arcade. In this case, 5 represents the price of admission to the arcade, and 2 represents the cost of each game played. To calculate the total cost, use the number of games played for x, multiply it by 2, and add 5.

Word Problems and Applications

In word problems, multiple quantities are often provided with a request to find some kind of relation between them. This often will mean that one variable (the dependent variable whose value needs to be found) can be written as a function of another variable (the independent variable whose value can be figured from the given information). The usual procedure for solving these problems is to start by giving each quantity in the problem a variable, and then figuring the relationship between these variables.

For example, suppose a car gets 25 miles per gallon. How far will the car travel if it uses 2.4 gallons of fuel? In this case, y would be the distance the car has traveled in miles, and x would be the amount of fuel burned in gallons (2.4). Then the relationship between these variables can be written as an algebraic equation, $y = 25x$. In this case, the equation is $y = 25 \times 2.4 = 60$, so the car has traveled 60 miles.

Some word problems require more than just one simple equation to be written and solved. Consider the following situations and the linear equations used to model them.

Suppose Margaret is 2 miles to the east of John at noon. Margaret walks to the east at 3 miles per hour. How far apart will they be at 3 p.m.? To solve this, x would represent the time in hours past noon, and y would represent the distance between Margaret and John. Now, noon corresponds to the equation where x is 0, so the y intercept is going to be 2. It's also known that the slope will be the rate at which the distance is changing, which is 3 miles per hour. This means that the slope will be 3 (be careful at this point: if units were used, other than miles and hours, for x and y variables, a conversion of the given information to the appropriate units would be required first). The simplest way to write an equation given the y-intercept, and the slope is the Slope-Intercept form, is $y = mx + b$. Recall that m here is the slope and b is the y intercept. So, $m = 3$ and $b = 2$. Therefore, the equation will be $y = 3x + 2$. The word problem asks how far to the east Margaret will be from John at 3 p.m., which means when x is 3. So, substitute $x = 3$ into this equation to obtain $y = 3 \times 3 + 2 = 9 + 2 = 11$. Therefore, she will be 11 miles to the east of him at 3 p.m.

For another example, suppose that a box with 4 cans in it weighs 6 lbs., while a box with 8 cans in it weighs 12 lbs. Find out how much a single can weighs. To do this, let x denote the number of cans in the box, and y denote the weight of the box with the cans in lbs. This line touches two pairs: (4, 6) and (8, 12). A formula for this relation could be written using the two-point form, with $x_1 = 4, y_1 = 6, x_2 = 8, y_2 = 12$. This would yield $\frac{y-6}{x-4} = \frac{12-6}{8-4}$, or $\frac{y-6}{x-4} = \frac{6}{4} = \frac{3}{2}$. However, only the slope is needed to solve this problem, since the slope will be the weight of a single can. From the computation, the slope is $\frac{3}{2}$. Therefore, each can weighs $\frac{3}{2}$ lb.

Intermediate Algebra and Functions

Polynomials

An expression of the form ax^n, where n is a non-negative integer, is called a *monomial* because it contains one term. A sum of monomials is called a *polynomial*. For example, $-4x^3 + x$ is a polynomial, while $5x^7$ is a monomial. A function equal to a polynomial is called a *polynomial function*.

The monomials in a polynomial are also called the *terms* of the polynomial.

The constants that precede the variables are called *coefficients*.

The highest value of the exponent of x in a polynomial is called the *degree* of the polynomial. So, $-4x^3 + x$ has a degree of 3, while $-2x^5 + x^3 + 4x + 1$ has a degree of 5. When multiplying polynomials, the degree of the result will be the sum of the degrees of the two polynomials being multiplied.

To add polynomials, add the coefficients of like powers of x. For example, $(-2x^5 + x^3 + 4x + 1) + (-4x^3 + x) = -2x^5 + (1 - 4)x^3 + (4 + 1)x + 1 = -2x^5 - 3x^3 + 5x + 1$.

Likewise, subtraction of polynomials is performed by subtracting coefficients of like powers of x. So, $(-2x^5 + x^3 + 4x + 1) - (-4x^3 + x) = -2x^5 + (1 + 4)x^3 + (4 - 1)x + 1 = -2x^5 + 5x^3 + 3x + 1$.

To multiply two polynomials, multiply each term of the first polynomial by each term of the second polynomial and add the results. For example, $(4x^2 + x)(-x^3 + x) = 4x^2(-x^3) + 4x^2(x) + x(-x^3) +$

$x(x) = -4x^5 + 4x^3 - x^4 + x^2$. In the case where each polynomial has two terms, like in this example, some students find it helpful to remember this as multiplying the First terms, then the Outer terms, then the Inner terms, and finally the Last terms, with the mnemonic FOIL. For longer polynomials, the multiplication process is the same, but there will be, of course, more terms, and there is no common mnemonic to remember each combination.

Factors for polynomials are similar to factors for integers—they are numbers, variables, or polynomials that, when multiplied together, give a product equal to the polynomial in question. One polynomial is a factor of a second polynomial if the second polynomial can be obtained from the first by multiplying by a third polynomial.

$6x^6 + 13x^4 + 6x^2$ can be obtained by multiplying together $(3x^4 + 2x^2)(2x^2 + 3)$. This means $2x^2 + 3$ and $3x^4 + 2x^2$ are factors of $6x^6 + 13x^4 + 6x^2$.

In general, finding the factors of a polynomial can be tricky. However, there are a few types of polynomials that can be factored in a straightforward way.

If a certain monomial is in each term of a polynomial, it can be factored out. There are several common forms polynomials take, which if you recognize, you can solve. The first example is a perfect square trinomial. To factor this polynomial, first expand the middle term of the expression:

$$x^2 + 2xy + y^2$$

$$x^2 + xy + xy + y^2$$

Factor out a common term in each half of the expression (in this case x from the left and y from the right):

$$x(x + y) + y(x + y)$$

Then the same can be done again, treating $(x + y)$ as the common factor:

$$(x + y)(x + y) = (x + y)^2$$

Therefore, the formula for this polynomial is:

$$x^2 + 2xy + y^2 = (x + y)^2$$

Next is another example of a perfect square trinomial. The process is the similar, but notice the difference in sign:

$$x^2 - 2xy + y^2$$

$$x^2 - xy - xy + y^2$$

Factor out the common term on each side:

$$x(x - y) - y(x - y)$$

Factoring out the common term again:

$$(x - y)(x - y) = (x - y)^2$$

Thus:

$$x^2 - 2xy + y^2 = (x - y)^2$$

The next is known as a difference of squares. This process is effectively the reverse of binomial multiplication:

$$x^2 - y^2$$

$$x^2 - xy + xy - y^2$$

$$x(x - y) + y(x - y)$$

$$(x + y)(x - y)$$

Therefore:

$$x^2 - y^2 = (x + y)(x - y)$$

The following two polynomials are known as the sum or difference of cubes. These are special polynomials that take the form of $x^3 + y^3$ or $x^3 - y^3$. The following formula factors the sum of cubes:

$$x^3 + y^3 = (x + y)(x^2 - xy + y^2)$$

Next is the difference of cubes, but note the change in sign. The formulas for both are similar, but the order of signs for factoring the sum or difference of cubes can be remembered by using the acronym SOAP, which stands for "same, opposite, always positive." The first sign is the same as the sign in the first expression, the second is opposite, and the third is always positive. The next formula factors the difference of cubes:

$$x^3 - y^3 = (x - y)(x^2 + xy + y^2)$$

The following two examples are expansions of cubed binomials. Similarly, these polynomials always follow a pattern:

$$x^3 + 3x^2y + 3xy^2 + y^3 = (x + y)^3$$

$$x^3 - 3x^2y + 3xy^2 - y^3 = (x - y)^3$$

These rules can be used in many combinations with one another. For example, the expression $3x^3 - 24$ has a common factor of 3, which becomes:

$$3(x^3 - 8)$$

A difference of cubes still remains which can then be factored out:

$$3(x - 2)(x^2 + 2x + 4)$$

There are no other terms to be pulled out, so this expression is completely factored.

When factoring polynomials, a good strategy is to multiply the factors to check the result. Let's try another example:

$$4x^3 + 16x^2$$

Both sides of the expression can be divided by 4, and both contain x^2, because $4x^3$ can be thought of as $4x^2(x)$, so the common term can simply be factored out:

$$4x^2(x + 4)$$

It sometimes can be necessary to rewrite the polynomial in some clever way before applying the above rules. Consider the problem of factoring $x^4 - 1$. This does not immediately look like any of the previous polynomials. However, it's possible to think of this polynomial as $x^4 - 1 = (x^2)^2 - (1^2)^2$, and now it can be treated as a difference of squares to simplify this:

$$(x^2)^2 - (1^2)^2$$

$$(x^2)^2 - x^2 1^2 + x^2 1^2 - (1^2)^2$$

$$x^2(x^2 - 1^2) + 1^2(x^2 - 1^2)$$

$$(x^2 + 1^2)(x^2 - 1^2)$$

$$(x^2 + 1)(x^2 - 1)$$

Quadratic Functions

A polynomial of degree 2 is called *quadratic*. Every quadratic function can be written in the form $ax^2 + bx + c$. The graph of a quadratic function, $y = ax^2 + bx + c$, is called a *parabola*. Parabolas are vaguely U-shaped.

Parabola

Whether the parabola opens upward or downward depends on the sign of *a*. If *a* is positive, then the parabola will open upward. If *a* is negative, then the parabola will open downward. The value of *a* will also affect how wide the parabola is. If the absolute value of *a* is large, then the parabola will be fairly skinny. If the absolute value of *a* is small, then the parabola will be quite wide.

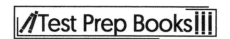

Changes to the value of b affect the parabola in different ways, depending on the sign of a. For positive values of a, increasing b will move the parabola to the left, and decreasing b will move the parabola to the right. On the other hand, if a is negative, the effects will be the opposite: increasing b will move the parabola to the right, while decreasing b will move the parabola to the left.

Changes to the value of c move the parabola vertically. The larger that c is, the higher the parabola gets. This does not depend on the value of a.

The quantity $D = b^2 - 4ac$ is called the *discriminant* of the parabola. When the discriminant is positive, then the parabola has two real zeros, or x intercepts. However, if the discriminant is negative, then there are no real zeros, and the parabola will not cross the x-axis. The highest or lowest point of the parabola is called the *vertex*. If the discriminant is zero, then the parabola's highest or lowest point is on the x-axis, and it will have a single real zero. The x-coordinate of the vertex can be found using the equation $x = -\frac{b}{2a}$. Plug this x-value into the equation and find the y-coordinate.

A quadratic equation is often used to model the path of an object thrown into the air. The x-value can represent the time in the air, while the y-value can represent the height of the object. In this case, the maximum height of the object would be the y-value found when the x-value is $-\frac{b}{2a}$.

Solving Quadratic Equations

A *quadratic equation* is an equation in the form $ax^2 + bx + c = 0$. There are several methods to solve such equations. The easiest method will depend on the quadratic equation in question.

It sometimes is possible to solve quadratic equations by manually *factoring* them. This means rewriting them in the form $(x + A)(x + B) = 0$. If this is done, then they can be solved by remembering that when $ab = 0$, either a or b must be equal to zero. Therefore, to have $(x + A)(x + B) = 0$, $(x + A) = 0$ or $(x + B) = 0$ is needed. These equations have the solutions $x = -A$ and $x = -B$, respectively.

In order to factor a quadratic equation, note that $(x + A)(x + B) = x^2 + (A + B)x + AB$. So, if an equation is in the form $x^2 + bx + c$, two numbers, A and B, need to be found that will add up to give us b, and multiply together to give us c.

As an example, consider solving the equation $-3x^2 + 6x + 9 = 0$. Start by dividing both sides by -3, leaving $x^2 - 2x - 3 = 0$. Now, notice that $1 - 3 = -2$, and also that $(1)(-3) = -3$. This means the equation can be factored into $(x + 1)(x - 3) = 0$. Now, solve $(x + 1) = 0$ and $(x - 3) = 0$ to get $x = -1$ and $x = 3$ as the solutions.

It is useful when trying to factor to remember that $x^2 + 2xy + y^2 = (x + y)^2$, $x^2 - 2xy + y^2 = (x - y)^2$, and $x^2 - y^2 = (x + y)(x - y)$.

However, factoring by hand is often hard to do. If there are no obvious ways to factor the quadratic equation, solutions can still be found by using the *quadratic formula*.

The quadratic formula is $x = \frac{-b \pm \sqrt{b^2 - 4ac}}{2a}$. This method will always work, although it sometimes can take longer than factoring by hand, if the factors are easy to guess. Using the standard form $ax^2 + bx + c = 0$, plug the values of a, b, and c from the equation into the formula and solve for x. There will either be two answers, one answer, or no real answer. No real answer comes when the value of the discriminant, the number under the square root, is a negative number. Since there are no real numbers that square to get a negative, the answer will be no real roots.

Here is an example of solving a quadratic equation using the quadratic formula. Suppose the equation to solve is $-2x^2 + 3x + 1 = 0$. There is no obvious way to factor this, so the quadratic formula is used, with $a = -2, b = 3, c = 1$. After substituting these values into the quadratic formula, it yields this:

$$x = \frac{-3 \pm \sqrt{3^2 - 4(-2)(1)}}{2(-2)}$$

This can be simplified to obtain:

$$\frac{3 \pm \sqrt{9 + 8}}{4}$$

or

$$\frac{3 \pm \sqrt{17}}{4}$$

Challenges can be encountered when asked to find a quadratic equation with specific roots. Given roots A and B, a quadratic function can be constructed with those roots by taking $(x - A)(x - B)$. So, in constructing a quadratic equation with roots $x = -2, 3$, it would result in $(x + 2)(x - 3) = x^2 - x - 6$. Multiplying this by a constant also could be done without changing the roots.

Exponents and Roots

An *exponent* is written as a^b. In this expression, a is called the *base* and b is called the *exponent*. It is properly stated that a is raised to the n-th power. Therefore, in the expression 2^3, the exponent is 3, while the base is 2. Such an expression is called an *exponential expression*. Note that when the exponent is 2, it is called *squaring* the base, and when it is 3, it is called *cubing* the base.

When the exponent is a positive integer, this indicates the base is multiplied by itself the number of times written in the exponent. So, in the expression 2^3, multiply 2 by itself with 3 copies of 2: $2^3 = 2 \times 2 \times 2 = 8$. One thing to notice is that, for positive integers n and m, $a^n a^m = a^{n+m}$ is a rule. In order to make this rule be true for an integer of 0, $a^0 = 1$, so that $a^n a^0 = a^{n+0} = a^n$. And, in order to make this rule be true for negative exponents, $a^{-n} = \frac{1}{a^n}$.

Another rule for simplifying expressions with exponents is shown by the following equation: $(a^m)^n = a^{mn}$. This is true for fractional exponents as well. So, for a positive integer, define $a^{\frac{1}{n}}$ to be the number that, when raised to the n-th power, provides a. In other words, $(a^{\frac{1}{n}})^n = a$ is the desired equation. It should be noted that $a^{\frac{1}{n}}$ is the n-th root of a. This also can be written as $a^{\frac{1}{n}} = \sqrt[n]{a}$. The symbol on the right-hand side of this equation is called a *radical*. If the root is left out, assume that the 2nd root should be taken, also called the *square* root: $a^{\frac{1}{2}} = \sqrt[2]{a} = \sqrt{a}$. Additionally, $\sqrt[3]{a}$ is also called the *cube* root.

Note that when multiple roots exist, $a^{\frac{1}{n}}$ is defined to be the *positive* root. So, $4^{\frac{1}{2}} = 2$. Also note that negative numbers do not have even roots in the real numbers.

This also enables finding exponents for any rational number: $a^{\frac{m}{n}} = (a^{\frac{1}{n}})^m = (a^m)^{\frac{1}{n}}$. In fact, the exponent can be any real number. In general, the following rules for exponents should be used for any numbers $a, b, m,$ and n.

- $a^1 = a$.
- $1^a = 1$.
- $a^0 = 1$.
- $a^m a^n = a^{m+n}$.
- $\frac{a^m}{a^n} = a^{m-n}$.
- $(a^m)^n = a^{m \times n}$.
- $(ab)^m = a^m b^m$.
- $\left(\frac{a}{b}\right)^m = \frac{a^m}{b^m}$.

As an example of applying these rules, consider the problem of simplifying the expression $(3x^2 y)^3 (2xy^4)$. Start by simplifying the left term using the sixth rule listed. Applying this rule yields the following expression: $27x^6 y^3 (2xy^4)$. The exponents can now be combined with base x and the exponents with base y. Multiply the coefficients to yield $54x^7 y^7$.

Order of Operations

When working with complicated expressions, parentheses are used to indicate in which order to perform operations. However, to avoid having too many parentheses in an expression, here are some basic rules concerning the proper order to perform operations when not otherwise specified.

1. Parentheses: always perform operations inside parentheses first, regardless of what those operations are
2. Exponents
3. Multiplication and Division
4. Addition and Subtraction

For #3 & #4, work these from left to right. So, if there a subtraction problem and then an addition problem, the subtraction problem will be worked first.

Note that multiplication and division are performed from left to right as they appear in the expression or equation. Addition and subtraction also are performed from left to right as they appear.

As an aid to memorizing this, some students like to use the mnemonic PEMDAS. Furthermore, this acronym can be associated with a mnemonic phrase such as "Pirates Eat Many Donuts At Sea."

Solving Equations with Exponents and Roots

Here are some of the most important properties of exponents and roots: if *n* is an integer, and if $a^n = b^n$, then $a = b$ if *n* is odd; but $a = \pm b$ if *n* is even. Similarly, if the roots of two things are equal, $\sqrt[n]{a} = \sqrt[n]{b}$, then $a = b$. This means that when starting with a true equation, both sides of that equation can be raised to a given power to obtain another true equation. Beware that when an even-powered root is taken on both sides of the equation, a $\pm$ in the result. For example, given the equation $x^2 = 16$, take the square root of both sides to solve for x. This results in the answer $x = \pm 4$ because $(-4)^2 = 16$ and $(4)^2 = 16$.

Another property is that if $a^n = a^m$, then $n = m$. This is true for any real numbers n and m.

For solving the equation $\sqrt{x+2} - 1 = 3$, start by moving the -1 over to the right-hand side. This is performed by adding 1 to both sides, which yields $\sqrt{x+2} = 4$. Now, square both sides, but remember that by squaring both sides, the signs are irrelevant. This yields $x + 2 = 16$, which simplifies to give $x = 14$.

Now consider the problem $(x+1)^4 = 16$. To solve this, take the 4$^\text{th}$ root of both sides, which means an ambiguity in the sign will be introduced because it is an even root: $\sqrt[4]{(x+1)^4} = \pm\sqrt[4]{16}$. The right-hand side is 2, since $2^4 = 16$. Therefore, $x + 1 = \pm 2$ or $x = -1 \pm 2$. Thus, the two possible solutions are $x = -3$ and $x = 1$.

Remember that when solving equations, the answer can be checked by plugging the solution back into the problem to make a true statement.

Rational Expressions

A rational expression is a fraction where the numerator and denominator are both polynomials. Some examples of rational expressions include the following: $\frac{4x^3y^5}{3z^4}$, $\frac{4x^3+3x}{x^2}$, and $\frac{x^2+7x+10}{x+2}$. Since these refer to expressions and not equations, they can be simplified but not solved. Using the rules in the previous Exponents and Roots sections, some rational expressions with monomials can be simplified. Other rational expressions such as the last example, $\frac{x^2+7x+10}{x+2}$, take more steps to be simplified. First, the polynomial on top can be factored from $x^2 + 7x + 10$ into $(x+5)(x+2)$. Then the common factors can be canceled and the expression can be simplified to $(x+5)$.

Consider this problem as an example of using rational expressions. Reggie wants to lay sod in his rectangular backyard. The length of the yard is given by the expression $4x + 2$ and the width is unknown. The area of the yard is $20x + 10$. Reggie needs to find the width of the yard. Knowing that the area of a rectangle is length multiplied by width, an expression can be written to find the width: $\frac{20x+10}{4x+2}$, area divided by length. Simplifying this expression by factoring out 10 on the top and 2 on the bottom leads to this expression: $\frac{10(2x+1)}{2(2x+1)}$. By cancelling out the $2x + 1$, that results in $\frac{10}{2} = 5$. The width of the yard is found to be 5 by simplifying a rational expression.

Rational Equations

A *rational equation* can be as simple as an equation with a ratio of polynomials, $\frac{p(x)}{q(x)}$, set equal to a value, where $p(x)$ and $q(x)$ are both polynomials. Notice that a rational equation has an equal sign, which is different from expressions. This leads to solutions, or numbers that make the equation true.

It is possible to solve rational equations by trying to get all of the x terms out of the denominator and then isolate them on one side of the equation. For example, to solve the equation $\frac{3x+2}{2x+3} = 4$, start by multiplying both sides by $(2x+3)$. This will cancel on the left side to yield $3x + 2 = 4(2x + 3)$, then $3x + 2 = 8x + 12$. Now, subtract $8x$ from both sides, which yields $-5x + 2 = 12$. Subtracting 2 from both sides results in $-5x = 10$. Finally, divide both sides by -5 to obtain $x = -2$.

Sometimes, when solving rational equations, it can be easier to try to simplify the rational expression by factoring the numerator and denominator first, then cancelling out common factors. For example, to solve $\frac{2x^2-8x+6}{x^2-3x+2} = 1$, start by factoring $2x^2 - 8x + 6 = 2(x^2 - 4x + 3) = 2(x - 1)(x - 3)$. Then, factor $x^2 - 3x + 2$ into $(x - 1)(x - 2)$. This turns the original equation into $\frac{2(x-1)(x-3)}{(x-1)(x-2)} = 1$. The common factor of $(x - 1)$ can be canceled, leaving us with $\frac{2(x-3)}{x-2} = 1$. Now the same method used in the previous example can be followed. Multiplying both sides by $x - 2$ and performing the multiplication on the left yields $2x - 6 = x - 2$, which can be simplified to $x = 4$.

Rational Functions

A rational function is similar to an equation, but it includes two variables. In general, a rational function is in the form: $f(x) = \frac{p(x)}{q(x)}$, where $p(x)$ and $q(x)$ are polynomials. Refer to the previous Functions section for a more detailed definition of functions. Rational functions are defined everywhere except where the denominator is equal to zero. When the denominator is equal to zero, this indicates either a hole in the graph or an asymptote. An asymptote can be either vertical, horizontal, or slant. A hole occurs when both the numerator and denominator are equal to 0 for a given value of x. A rational function can have at most one vertical asymptote and one horizontal or slant asymptote. An asymptote is a line such that the distance between the curve and the line tends toward 0, but never reaches it, as the line heads toward infinity. Examples of these types of functions are shown below. The first graph shows a rational function with a vertical asymptote at x = 0. This can be found by setting the denominator equal to 0. In this case it is just x = 0.

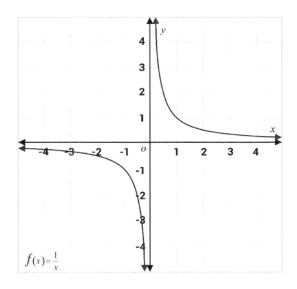

$f(x) = \frac{1}{x}$

The second graph shows a rational function with a vertical asymptote at x = -.5. Again this can be found by just setting the denominator equal to 0. So, $2x^2 + x = 0, 2x + 1 = 0, 2x = -1, x = -.5$. This graph also has a hole in the graph at $x = 0$. This is because both the numerator and denominator are equal to 0 when $x = 0$.

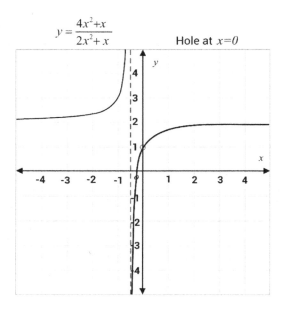

$$y = \frac{4x^2+x}{2x^2+x}$$ Hole at $x=0$

Geometry and Measurement

Plane Geometry

Locations on the plane that have no width or breadth are called *points*. These points usually will be denoted with capital letters such as P.

Any pair of points A, B on the plane will determine a unique straight line between them. This line is denoted AB. Sometimes to emphasize a line is being considered, this will be written as $\overleftrightarrow{AB}$.

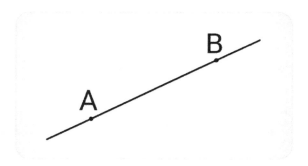

If the Cartesian coordinates for A and B are known, then the distance $d(A, B)$ along the line between them can be measured using the *Pythagorean formula*, which states that if $A = (x_1, y_1)$ and $B = (x_2, y_2)$, then the distance between them is $d(A, B) = \sqrt{(x_2 - x_1)^2 + (y_2 - y_1)^2}$.

The part of a line that lies between *A* and *B* is called a *line segment*. It has two endpoints, one at *A* and one at *B*. *Rays* also can be formed. Given points *A* and *B*, a *ray* is the portion of a line that starts at one of these points, passes through the other, and keeps on going. Therefore, a ray has a single endpoint, but the other end goes off to infinity.

Given a pair of points *A* and *B*, a circle centered at *A* and passing through *B* can be formed. This is the set of points whose distance from *A* is exactly $d(A, B)$. The radius of this circle will be $d(A, B)$.

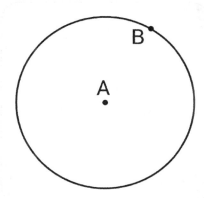

The *circumference* of a circle is the distance traveled by following the edge of the circle for one complete revolution, and the length of the circumference is given by $2\pi r$, where *r* is the radius of the circle. The formula for circumference is $C = 2\pi r$.

When two lines cross, they form an *angle*. The point where the lines cross is called the *vertex* of the angle. The angle can be named by either just using the vertex, $\angle A$, or else by listing three points $\angle BAC$, as shown in the diagram below.

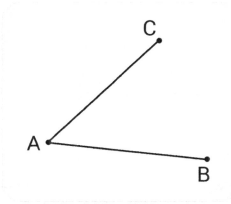

The measurement of an angle can be given in degrees or in radians. In degrees, a full circle is 360 degrees, written 360°. In radians, a full circle is 2π radians.

Given two points on the circumference of a circle, the path along the circle between those points is called an *arc* of the circle. For example, the arc between *B* and *C* is denoted by a thinner line:

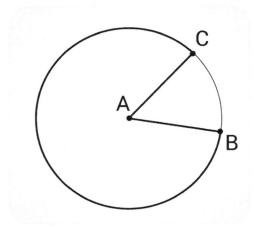

The length of the path along an arc is called the *arc length*. If the circle has radius *r*, then the arc length is given by multiplying the measure of the angle in radians by the radius of the circle.

Two lines are said to be *parallel* if they never intersect. If the lines are *AB* and *CD*, then this is written as *AB* ∥ *CD*.

If two lines cross to form four quarter-circles, that is, 90° angles, the two lines are *perpendicular*. If the point at which they cross is *B*, and the two lines are *AB* and *BC*, then this is written as *AB* ⊥ *BC*.

A *polygon* is a closed figure (meaning it divides the plane into an inside and an outside) consisting of a collection of line segments between points. These points are called the *vertices* of the polygon. These line segments must not overlap one another. Note that the number of sides is equal to the number of angles, or vertices of the polygon. The angles between line segments meeting one another in the polygon are called *interior angles*.

A *regular polygon* is a polygon whose edges are all the same length and whose interior angles are all of equal measure.

A *triangle* is a polygon with three sides. A *quadrilateral* is a polygon with four sides.

A *right triangle* is a triangle that has one 90° angle.

The sum of the interior angles of any triangle must add up to 180°.

An *isosceles triangle* is a triangle in which two of the sides are the same length. In this case, it will always have two congruent interior angles. If a triangle has two congruent interior angles, it will always be isosceles.

An *equilateral triangle* is a triangle whose sides are all the same length and whose angles are all equivalent to one another, equal to 60°. Equilateral triangles are examples of regular polygons. Note that equilateral triangles are also isosceles.

A *rectangle* is a quadrilateral whose interior angles are all 90°. A rectangle has two sets of sides that are equal to one another.

A *square* is a rectangle whose width and height are equal. Therefore, squares are regular polygons.

A *parallelogram* is a quadrilateral in which the opposite sides are parallel and equivalent to each other.

Transformations of a Plane

Given a figure drawn on a plane, many changes can be made to that figure, including *rotation*, *translation*, and *reflection*. Rotations turn the figure about a point, translations slide the figure, and reflections flip the figure over a specified line. When performing these transformations, the original figure is called the *pre-image*, and the figure after transformation is called the *image*.

More specifically, *translation* means that all points in the figure are moved in the same direction by the same distance. In other words, the figure is slid in some fixed direction. Of course, while the entire figure is slid by the same distance, this does not change any of the measurements of the figures involved. The result will have the same distances and angles as the original figure.

In terms of Cartesian coordinates, a translation means a shift of each of the original points (x, y) by a fixed amount in the x and y directions, to become $(x + a, y + b)$.

Another procedure that can be performed is called *reflection*. To do this, a line in the plane is specified, called the *line of reflection*. Then, take each point and flip it over the line so that it is the same distance from the line but on the opposite side of it. This does not change any of the distances or angles involved, but it does reverse the order in which everything appears.

To reflect something over the x-axis, the points (x, y) are sent to $(x, -y)$. To reflect something over the y-axis, the points (x, y) are sent to the points $(-x, y)$. Flipping over other lines is not something easy to express in Cartesian coordinates. However, by drawing the figure and the line of reflection, the distance to the line and the original points can be used to find the reflected figure.

Example: Reflect this triangle with vertices (-1, 0), (2, 1), and (2, 0) over the *y*-axis. The pre-image is shown below.

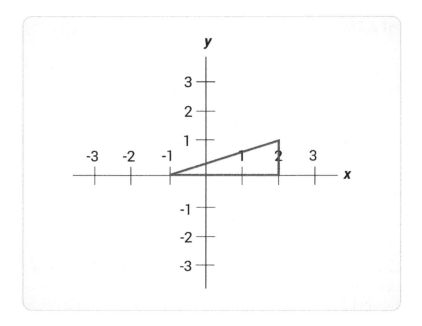

To do this, flip the *x* values of the points involved to the negatives of themselves, while keeping the *y* values the same. The image is shown here.

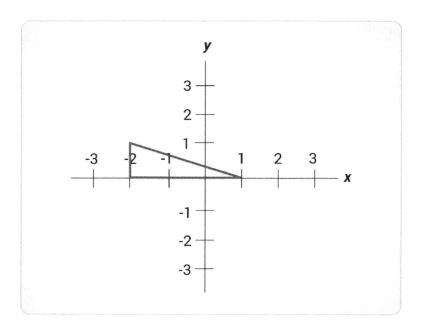

The new vertices will be (1, 0), (-2, 1), and (-2, 0).

Another procedure that does not change the distances and angles in a figure is *rotation*. In this procedure, pick a center point, then rotate every vertex along a circle around that point by the same

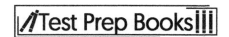

angle. This procedure is also not easy to express in Cartesian coordinates, and this is not a requirement on this test. However, as with reflections, it's helpful to draw the figures and see what the result of the rotation would look like. This transformation can be performed using a compass and protractor.

Each one of these transformations can be performed on the coordinate plane without changes to the original dimensions or angles.

If two figures in the plane involve the same distances and angles, they are called *congruent figures*. In other words, two figures are congruent when they go from one form to another through reflection, rotation, and translation, or a combination of these.

Remember that rotation and translation will give back a new figure that is identical to the original figure, but reflection will give back a mirror image of it.

To recognize that a figure has undergone a rotation, check to see that the figure has not been changed into a mirror image, but that its orientation has changed (that is, whether the parts of the figure now form different angles with the *x* and *y* axes).

To recognize that a figure has undergone a translation, check to see that the figure has not been changed into a mirror image, and that the orientation remains the same.

To recognize that a figure has undergone a reflection, check to see that the new figure is a mirror image of the old figure.

Keep in mind that sometimes a combination of translations, reflections, and rotations may be performed on a figure.

Dilation

A *dilation* is a transformation that preserves angles, but not distances. This can be thought of as stretching or shrinking a figure. If a dilation makes figures larger, it is called an *enlargement*. If a dilation makes figures smaller, it is called a *reduction*. The easiest example is to dilate around the origin. In this case, multiply the *x* and *y* coordinates by a *scale factor*, k, sending points (x, y) to (kx, ky).

As an example, draw a dilation of the following triangle, whose vertices will be the points (-1, 0), (1, 0), and (1, 1).

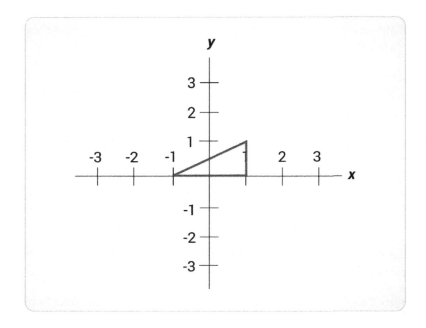

For this problem, dilate by a scale factor of 2, so the new vertices will be (-2, 0), (2, 0), and (2, 2).

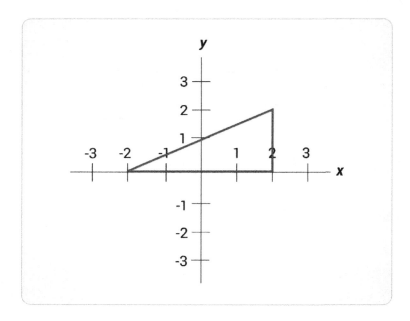

Note that after a dilation, the distances between the vertices of the figure will have changed, but the angles remain the same. The two figures that are obtained by dilation, along with possibly translation, rotation, and reflection, are all *similar* to one another. Another way to think of this is that similar figures have the same number of vertices and edges, and their angles are all the same. Similar figures have the same basic shape, but are different in size.

Symmetry

Using the types of transformations above, if an object can undergo these changes and not appear to have changed, then the figure is symmetrical. If an object can be split in half by a line and flipped over that line to lie directly on top of itself, it is said to have *line symmetry*. An example of both types of figures is seen below.

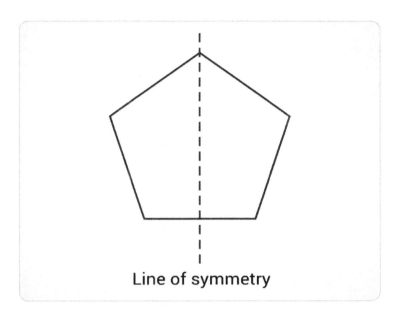

Line of symmetry

If an object can be rotated about its center to any degree smaller than 360, and it lies directly on top of itself, the object is said to have *rotational symmetry*. An example of this type of symmetry is shown below. The pentagon has an order of 5.

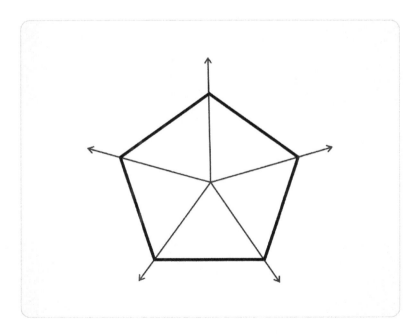

The rotational symmetry lines in the figure above can be used to find the angles formed at the center of the pentagon. Knowing that all of the angles together form a full circle, at 360 degrees, the figure can be split into 5 angles equally. By dividing the 360° by 5, each angle is 72°.

Given the length of one side of the figure, the perimeter of the pentagon can also be found using rotational symmetry. If one side length was 3 cm, that side length can be rotated onto each other side length four times. This would give a total of 5 side lengths equal to 3 cm. To find the perimeter, or distance around the figure, multiply 3 by 5. The perimeter of the figure would be 15 cm.

If a line cannot be drawn anywhere on the object to flip the figure onto itself or rotated less than or equal to 180 degrees to lay on top of itself, the object is asymmetrical. Examples of these types of figures are shown below.

Symmetric   Asymmetric

No line of symmetry

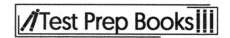

Perimeters and Areas

The *perimeter* of a polygon is the total length of a trip around the whole polygon, starting and ending at the same point. It is found by adding up the lengths of each line segment in the polygon. For a rectangle with sides of length x and y, the perimeter will be $2x + 2y$.

The area of a polygon is the area of the region that it encloses. Regarding the area of a rectangle with sides of length x and y, the area is given by xy. For a triangle with a base of length b and a height of length h, the area is $\frac{1}{2}bh$.

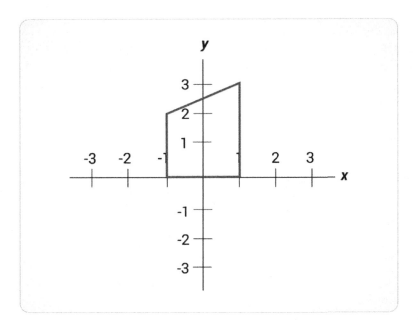

To find the areas of more general polygons, it is usually easiest to break up the polygon into rectangles and triangles. For example, find the area of the following figure whose vertices are (-1, 0), (-1, 2), (1, 3), and (1, 0).

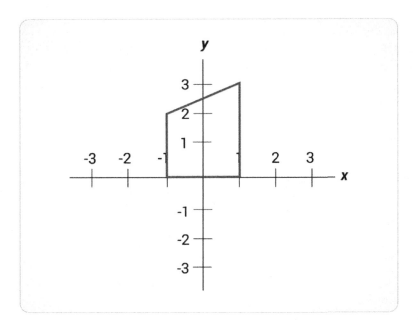

Separate this into a rectangle and a triangle as shown:

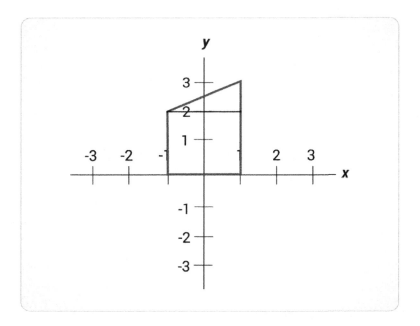

The rectangle has a height of 2 and a width of 2, so it has a total area of $2 \times 2 = 4$. The triangle has a width of 2 and a height of 1, so it has an area of $\frac{1}{2}2 \times 1 = 1$. Therefore, the entire quadrilateral has an area of 4 + 1 = 5.

As another example, suppose someone wants to tile a rectangular room that is 10 feet by 6 feet using triangular tiles that are 12 inches by 6 inches. How many tiles would be needed? To figure this, first find the area of the room, which will be $10 \times 6 = 60$ square feet. The dimensions of the triangle are 1 foot by ½ foot, so the area of each triangle is $\frac{1}{2} \times 1 \times \frac{1}{2} = \frac{1}{4}$ square feet. Notice that the dimensions of the triangle had to be converted to the same units as the rectangle. Now, take the total area divided by the area of one tile to find the answer: $\frac{60}{\frac{1}{4}} = 60 \times 4 = 240$ tiles required.

Volumes and Surface Areas

Geometry in three dimensions is similar to geometry in two dimensions. The main new feature is that three points now define a unique *plane* that passes through each of them. Three dimensional objects can be made by putting together two dimensional figures in different surfaces. Below, some of the possible three dimensional figures will be provided, along with formulas for their volumes and surface areas.

A rectangular prism is a box whose sides are all rectangles meeting at 90° angles. Such a box has three dimensions: length, width, and height. If the length is x, the width is y, and the height is z, then the volume is given by $V = xyz$.

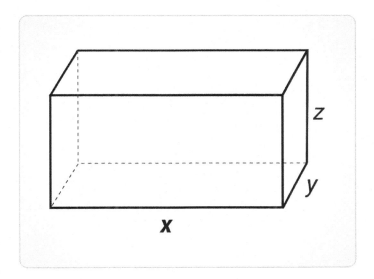

The surface area will be given by computing the surface area of each rectangle and adding them together. There are a total of six rectangles. Two of them have sides of length x and y, two have sides of length y and z, and two have sides of length x and z. Therefore, the total surface area will be given by $SA = 2xy + 2yz + 2xz$.

A *rectangular pyramid* is a figure with a rectangular base and four triangular sides that meet at a single vertex. If the rectangle has sides of length x and y, then the volume will be given by $V = \frac{1}{3}xyh$.

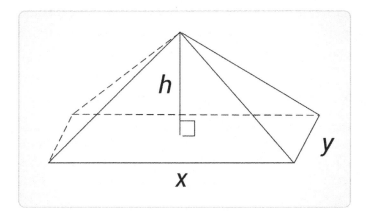

To find the surface area, the dimensions of each triangle need to be known. However, these dimensions can differ depending on the problem in question. Therefore, there is no general formula for calculating total surface area.

A *sphere* is a set of points all of which are equidistant from some central point. It is like a circle, but in three dimensions. The volume of a sphere of radius r is given by $V = \frac{4}{3}\pi r^3$. The surface area is given by $A = 4\pi r^2$.

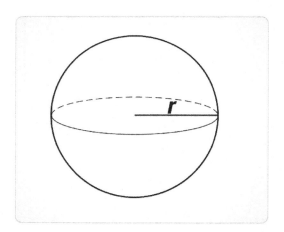

The Pythagorean Theorem

The Pythagorean theorem is an important result in geometry. It states that for right triangles, the sum of the squares of the two shorter sides will be equal to the square of the longest side (also called the *hypotenuse*). The longest side will always be the side opposite to the 90° angle. If this side is called c, and the other two sides are a and b, then the Pythagorean theorem states that $c^2 = a^2 + b^2$. Since lengths are always positive, this also can be written as $c = \sqrt{a^2 + b^2}$.

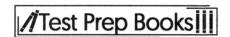

A diagram to show the parts of a triangle using the Pythagorean theorem is below.

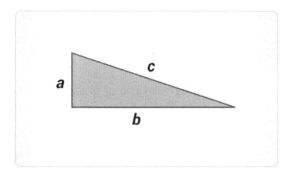

As an example of the theorem, suppose that Shirley has a rectangular field that is 5 feet wide and 12 feet long, and she wants to split it in half using a fence that goes from one corner to the opposite corner. How long will this fence need to be? To figure this out, note that this makes the field into two right triangles, whose hypotenuse will be the fence dividing it in half. Therefore, the fence length will be given by $\sqrt{5^2 + 12^2} = \sqrt{169} = 13$ feet long.

Similar Figures and Proportions

Sometimes, two figures are similar, meaning they have the same basic shape and the same interior angles, but they have different dimensions. If the ratio of two corresponding sides is known, then that ratio, or scale factor, holds true for all of the dimensions of the new figure.

Here is an example of applying this principle. Suppose that Lara is 5 feet tall and is standing 30 feet from the base of a light pole, and her shadow is 6 feet long. How high is the light on the pole? To figure this, it helps to make a sketch of the situation:

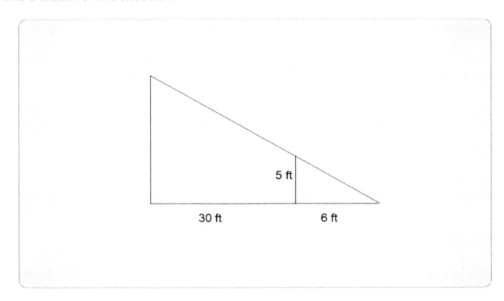

The light pole is the left side of the triangle. Lara is the 5-foot vertical line. Notice that there are two right triangles here, and that they have all the same angles as one another. Therefore, they form similar triangles. So, figure the ratio of proportionality between them.

The bases of these triangles are known. The small triangle, formed by Lara and her shadow, has a base of 6 feet. The large triangle, formed by the light pole along with the line from the base of the pole out to the end of Lara's shadow is $30 + 6 = 36$ feet long. So, the ratio of the big triangle to the little triangle will be $\frac{36}{6} = 6$. The height of the little triangle is 5 feet. Therefore, the height of the big triangle will be $6 \times 5 = 30$ feet, meaning that the light is 30 feet up the pole.

Notice that the perimeter of a figure changes by the ratio of proportionality between two similar figures, but the area changes by the *square* of the ratio. This is because if the length of one side is doubled, the area is quadrupled.

As an example, suppose two rectangles are similar, but the edges of the second rectangle are three times longer than the edges of the first rectangle. The area of the first rectangle is 10 square inches. How much more area does the second rectangle have than the first?

To answer this, note that the area of the second rectangle is $3^2 = 9$ times the area of the first rectangle, which is 10 square inches. Therefore, the area of the second rectangle is going to be $9 \times 10 = 90$ square inches. This means it has $90 - 10 = 80$ square inches more area than the first rectangle.

As a second example, suppose X and Y are similar right triangles. The hypotenuse of X is 4 inches. The area of Y is $\frac{1}{4}$ the area of X. What is the hypotenuse of Y?

First, realize the area has changed by a factor of $\frac{1}{4}$. The area changes by a factor that is the *square* of the ratio of changes in lengths, so the ratio of the lengths is the square root of the ratio of areas. That means that the ratio of lengths must be is $\sqrt{\frac{1}{4}} = \frac{1}{2}$, and the hypotenuse of Y must be $\frac{1}{2} \times 4 = 2$ inches.

Volumes between similar solids change like the cube of the change in the lengths of their edges. Likewise, if the ratio of the volumes between similar solids is known, the ratio between their lengths is known by finding the cube root of the ratio of their volumes.

For example, suppose there are two similar rectangular pyramids X and Y. The base of X is 1 inch by 2 inches, and the volume of X is 8 inches. The volume of Y is 64 inches. What are the dimensions of the base of Y?

To answer this, first find the ratio of the volume of Y to the volume of X. This will be given by $\frac{64}{8} = 8$. Now the ratio of lengths is the cube root of the ratio of volumes, or $\sqrt[3]{8} = 2$. So, the dimensions of the base of Y must be 2 inches by 4 inches.

Data Analysis, Statistics, and Probability

What are Statistics?

The field of statistics describes relationships between quantities that are related, but not necessarily in a deterministic manner. For example, a graduating student's salary will often be higher when the student graduates with a higher GPA, but this is not always the case. Likewise, people who smoke tobacco are more likely to develop lung cancer, but, in fact, it is possible for non-smokers to develop the disease as well. *Statistics* describes these kinds of situations, where the likelihood of some outcome depends on the starting data.

Descriptive statistics involves analyzing a collection of data to describe its broad properties such average (or mean), what percent of the data falls within a given range, and other such properties. An example of this would be taking all of the test scores from a given class and calculating the average test score. *Inferential statistics* attempts to use data about a subset of some population to make inferences about the rest of the population. An example of this would be taking a collection of students who received tutoring and comparing their results to a collection of students who did not receive tutoring, then using that comparison to try to predict whether the tutoring program in question is beneficial.

To be sure that inferences have a high probability of being true for the whole population, the subset that is analyzed needs to resemble a miniature version of the population as closely as possible. For this reason, statisticians like to choose random samples from the population to study, rather than picking a specific group of people based on some similarity. For example, studying the incomes of people who live in Portland does not tell anything useful about the incomes of people who live in Tallahassee.

Mean, Median, and Mode

Mean

Suppose that you have a set of data points and some description of the general properties of this data need to be found.

The first property that can be defined for this set of data is the *mean*. This is the same as average. To find the mean, add up all the data points, then divide by the total number of data points. For example, suppose that in a class of 10 students, the scores on a test were 50, 60, 65, 65, 75, 80, 85, 85, 90, 100. Therefore, the average test score will be:

$$\frac{50 + 60 + 65 + 65 + 75 + 80 + 85 + 85 + 90 + 100}{10} = 75.5$$

The mean is a useful number if the distribution of data is normal (more on this later), which roughly means that the frequency of different outcomes has a single peak and is roughly equally distributed on both sides of that peak. However, it is less useful in some cases where the data might be split or where there are some *outliers*. Outliers are data points that are far from the rest of the data. For example, suppose there are 10 executives and 90 employees at a company. The executives make $1000 per hour, and the employees make $10 per hour.

Therefore, the average pay rate will be:

$$\frac{\$1000 \times 10 + \$10 \times 90}{100} = \$109 \text{ per hour}$$

In this case, this average is not very descriptive since it's not close to the actual pay of the executives or the employees.

Median

Another useful measurement is the *median*. In a data set, the median is the point in the middle. The middle refers to the point where half the data comes before it and half comes after, when the data is recorded in numerical order. For instance, these are the speeds of the fastball of a pitcher during the last inning that he pitched (in order from least to greatest):

90, 92, 93, 93, 95, 96, 97, 97, 97

There are nine total numbers, so the middle or *median* number is the 5th one, which is 95.

In cases where the number of data points is an even number, then the average of the two middle points is taken. In the previous example of test scores, the two middle points are 75 and 80. Since there is no single point, the average of these two scores needs to be found. The average is:

$$\frac{75 + 80}{2} = 77.5$$

The median is generally a good value to use if there are a few outliers in the data. It prevents those outliers from affecting the "middle" value as much as when using the mean.

Since an outlier is a data point that is far from most of the other data points in a data set, this means an outlier also is any point that is far from the median of the data set. The outliers can have a substantial effect on the mean of a data set, but they usually do not change the median or mode, or do not change them by a large quantity. For example, consider the data set (3, 5, 6, 6, 6, 8). This has a median of 6 and a mode of 6, with a mean of $\frac{34}{6} \approx 5.67$. Now, suppose a new data point of 1000 is added so that the data set is now (3, 5, 6, 6, 6, 8, 1000). This does not change the median or mode, which are both still 6. However, the average is now $\frac{1034}{7}$, which is approximately 147.7. In this case, the median and mode will be better descriptions for most of the data points.

The reason for outliers in a given data set is a complicated problem. It is sometimes the result of an error by the experimenter, but often they are perfectly valid data points that must be taken into consideration.

Mode

One additional measure to define for *X* is the *mode*. This is the data point that appears most frequently. If two or more data points all tie for the most frequent appearance, then each of them is considered a mode. In the case of the test scores, where the numbers were 50, 60, 65, 65, 75, 80, 85, 85, 90, 100, there are two modes: 65 and 85.

Quartiles and Percentiles

The *first quartile* of a set of data *X* refers to the largest value from the first ¼ of the data points. In practice, there are sometimes slightly different definitions that can be used, such as the median of the first half of the data points (excluding the median itself if there are an odd number of data points). The term also has a slightly different use: when it is said that a data point lies *in the first quartile*, it means it is less than or equal to the median of the first half of the data points. Conversely, if it lies *at* the first quartile, then it is equal to the first quartile.

When it is said that a data point lies in the *second quartile*, it means it is between the first quartile and the median.

The *third quartile* refers to data that lies between ½ and ¾ of the way through the data set. Again, there are various methods for defining this precisely, but the simplest way is to include all of the data that lie between the median and the median of the top half of the data.

Data that lies in the *fourth quartile* refers to all of the data above the third quartile.

Percentiles may be defined in a similar manner to quartiles. Generally, this is defined in the following manner:

If a data point lies *in the n-th percentile*, this means it lies in the range of the first *n%* of the data.

If a data point lies *at the n-th percentile*, then it means that *n%* of the data lies below this data point.

Standard Deviation

Given a data set X consisting of data points $(x_1, x_2, x_3, \ldots x_n)$, the *variance* of X is defined to be $\frac{\sum_{i=1}^{n}(x_i - \bar{X})^2}{n}$. This means that the variance of X is the average of the squares of the differences between each data point and the mean of X. In the formula, $\bar{X}$ is the mean of the values in the data set, and x_i represents each individual value in the data set. The sigma notation indicates that the sum should be found with n being the number of values to add together. $i = 1$ means that the values should begin with the first value.

Given a data set X consisting of data points $(x_1, x_2, x_3, \ldots x_n)$, the *standard deviation* of X is defined to be $s_x = \sqrt{\frac{\sum_{i=1}^{n}(x_i - \bar{X})^2}{n}}$. In other words, the standard deviation is the square root of the variance.

Both the variance and the standard deviation are measures of how much the data tend to be spread out. When the standard deviation is low, the data points are mostly clustered around the mean. When the standard deviation is high, this generally indicates that the data are quite spread out, or else that there are a few substantial outliers.

As a simple example, compute the standard deviation for the data set (1, 3, 3, 5). First, compute the mean, which will be $\frac{1+3+3+5}{4} = \frac{12}{4} = 3$. Now, find the variance of X with the formula:

$$\sum_{i=1}^{4}(x_i - \bar{X})^2 = (1-3)^2 + (3-3)^2 + (3-3)^2 + (5-3)^2$$

$$-2^2 + 0^2 + 0^2 + 2^2 = 8$$

Therefore, the variance is $\frac{8}{4} = 2$. Taking the square root, the standard deviation will be $\sqrt{2}$.

Note that the standard deviation only depends upon the mean, not upon the median or mode(s). Generally, if there are multiple modes that are far apart from one another, the standard deviation will be high. A high standard deviation does not always mean there are multiple modes, however.

Fitting Functions to Data

Sometimes, when data are measured, it is not simply measuring the frequency of a given outcome, but rather measuring a relationship between two different quantities. In these cases, there is usually one variable that is controlled, the *independent variable*, and one that depends on this variable, the *dependent variable*. If there is a relationship between the two variables, then they are said to be *correlated*.

There are two caveats to these terms. First, the independent variable is not necessarily controlled by the experimenters. It is simply the one chosen to organize the data. In other words, the data are divided up based on the value of an independent variable. Second, finding a significant relationship between the dependent variable and the independent variable does not necessarily imply that there is a causal relationship between the two variables. It only means that once the independent variable is known, a fairly accurate prediction of the dependent variable can be made. This is often expressed with the phrase *correlation does not imply causation*. In other words, just because there is a relationship between two variables does not mean that one is the cause of the other. There could be other factors involved that are the real cause.

Consider some examples. An experimenter could do an experiment in which the independent variable is the number of hours that a student studies for a given test, and the dependent variable is the score the student receives when he or she actually takes the test. Such an experiment would attempt to measure whether there is a relationship between the time spent studying and the score a student receives when taking the test. The expectation would be that the larger value of the independent variable would yield a larger value for the dependent variable. Another experimenter might do an experiment with runners, where the independent variable is the length of the runner's leg, and the dependent variable is the time it takes for the runner to run a fixed distance. In this experiment, as the independent variable increases, the dependent variable would be expected to decrease.

As an example of the phenomenon that correlation does not imply causation, consider an experiment where the independent variable is the value of a person's house, and the dependent variable is their income. Although people in more expensive houses are expected to make more money, it is clear that their expensive houses are not the cause of them making more money. This illustrates one example of why it is important for experimenters to be careful when drawing conclusions about causation from their data.

Linear Data Fitting

The simplest type of correlation between two variables is a *linear correlation*. If the independent variable is x and the dependent variable is y, then a linear correlation means $y = mx + b$. If m is positive, then y will increase as x increases. While if m is negative, then y decreases while x increases. The variable b represents the value of y when x is 0.

As one example of such a correlation, consider a manufacturing plant. Suppose x is the number of units produced by the plant, and y is the cost to the company. In this example, b will be the cost of the plant itself. The plant will cost money even if it is never used, just by buying the machinery. For each unit produced, there will be a cost for the labor and the material. Let m represent this cost to produce one unit of the product.

For a more concrete example, suppose a computer factory costs $100,000. It requires $100 of parts and $50 of labor to make one computer. How much will it cost for a company to make 1000 computers? To figure this, let y be the amount of money the company spends, and let x be the number of computers. The cost of the factory is $100,000, so $b = 100,000$. On the other hand, the cost of producing a computer is the parts plus labor, or $150, so $m = 150$. Therefore, $y = 150x + 100,000$. Substitute 1000 for x and get $y = 150 \times 1000 + 100,000 = 150,000 + 1000 = 250,000$. It will cost the company $250,000 to make 1000 computers.

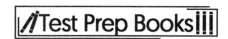

Probabilities

Given a set of possible outcomes X, a *probability distribution* on X is a function that assigns a probability to each possible outcome. If the outcomes are $(x_1, x_2, x_3, \ldots x_n)$, and the probability distribution is p, then the following rules are applied.

- $0 \leq p(x_i) \leq 1$, for any i.

- $\sum_{i=1}^{n} p(x_i) = 1$.

In other words, the probability of a given outcome must be between zero and 1, while the total probability must be 1.

If $p(x_i)$ is constant, then this is called a *uniform probability distribution*, and $p(x_i) = \frac{1}{n}$. For example, on a six-sided die, the probability of each of the six outcomes will be $\frac{1}{6}$.

If seeking the probability of an outcome occurring in some specific range A of possible outcomes, written $P(A)$, add up the probabilities for each outcome in that range. For example, consider a six-sided die, and figure the probability of getting a 3 or lower when it is rolled. The possible rolls are 1, 2, 3, 4, 5, and 6. So, to get a 3 or lower, a roll of 1, 2, or 3 must be completed. The probabilities of each of these is $\frac{1}{6}$, so add these to get $p(1) + p(2) + p(3) = \frac{1}{6} + \frac{1}{6} + \frac{1}{6} = \frac{1}{2}$.

Conditional Probabilities

An outcome occasionally lies within some range of possibilities B, and the probability that the outcomes also lie within some set of possibilities A needs to be figured. This is called a *conditional probability*. It is written as $P(A|B)$, which is read "the probability of A given B." The general formula for computing conditional probabilities is $P(A|B) = \frac{P(A \cap B)}{P(B)}$.

However, when dealing with uniform probability distributions, simplify this a bit. Write $|A|$ to indicate the number of outcomes in A. Then, for uniform probability distributions, write $P(A|B) = \frac{|A \cap B|}{|B|}$ (recall that $A \cap B$ means "A intersect B," and consists of all of the outcomes that lie in both A and B). This means that all possible outcomes do not need to be known. To see why this formula works, suppose that the set of outcomes X is $(x_1, x_2, x_3, \ldots x_n)$, so that $|X| = n$. Then, for a uniform probability distribution, $P(A) = \frac{|A|}{n}$. However, this means $(A|B) = \frac{P(A \cap B)}{P(B)} = \frac{\frac{|A \cap B|}{n}}{\frac{|B|}{n}} = \frac{|A \cap B|}{|B|}$, since the n's cancel out.

For example, suppose a die is rolled, and it is known that it will land between 1 and 4. However, how many sides the die has is unknown. Figure the probability that the die is rolled higher than 2. To figure this, $P(3)$ or $P(4)$ does not need to be determined, or any of the other probabilities, since it is known that a fair die has a uniform probability distribution. Therefore, apply the formula $\frac{|A \cap B|}{|B|}$. So, in this case B is (1, 2, 3, 4) and $A \cap B$ is (3, 4). Therefore $\frac{|A \cap B|}{|B|} = \frac{2}{4} = \frac{1}{2}$.

Conditional probability is an important concept because, in many situations, the likelihood of one outcome can differ radically depending on how something else comes out. The probability of passing a test given that one has studied all of the material is generally much higher than the probability of passing a test given that one has not studied at all. The probability of a person having heart trouble is much lower if that person exercises regularly. The probability that a college student will graduate is higher when his or her SAT scores are higher, and so on. For this reason, there are many people who are interested in conditional probabilities.

Note that in some practical situations, changing the order of the conditional probabilities can make the outcome very different. For example, the probability that a person with heart trouble has exercised regularly is quite different than the probability that a person who exercises regularly will have heart trouble. The probability of a person receiving a military-only award, given that he or she is or was a soldier, is generally not very high, but the probability that a person being or having been a soldier, given that he or she received a military-only award, is 1.

However, in some cases, the outcomes do not influence one another this way. If the probability of A is the same regardless of whether B is given; that is, if $P(A|B) = P(A)$, then A and B are considered *independent*. In this case, $P(A|B) = \frac{P(A \cap B)}{P(B)} = P(A)$, so $P(A \cap B) = P(A)P(B)$. In fact, if $P(A \cap B) = P(A)P(B)$, it can be determined that $P(A|B) = P(A)$ and $P(A|B) = P(B)$ by working backward. Therefore, B is also independent of A.

An example of something being independent can be seen in rolling dice. In this case, consider a red die and a green die. It is expected that when the dice are rolled, the outcome of the green die should not depend in any way on the outcome of the red die. Or, to take another example, if the same die is rolled repeatedly, then the next number rolled should not depend on which numbers have been rolled previously. Similarly, if a coin is flipped, then the next flip's outcome does not depend on the outcomes of previous flips.

This can sometimes be counter-intuitive, since when rolling a die or flipping a coin, there can be a streak of surprising results. If, however, it is known that the die or coin is fair, then these results are just the result of the fact that over long periods of time, it is very likely that some unlikely streaks of outcomes will occur. Therefore, avoid making the mistake of thinking that when considering a series of independent outcomes, a particular outcome is "due to happen" simply because a surprising series of outcomes has already been seen.

There is a second type of common mistake that people tend to make when reasoning about statistical outcomes: the idea that when something of low probability happens, this is surprising. It would be surprising that something with low probability happened after just one attempt. However, with so much happening all at once, it is easy to see at least something happen in a way that seems to have a very low probability. In fact, a lottery is a good example. The odds of winning a lottery are very small, but the odds that somebody wins the lottery each week are actually fairly high. Therefore, no one should be surprised when some low probability things happen.

Addition Rule

The *addition rule* for probabilities states that the probability of A or B happening is $P(A \cup B) = P(A) + P(B) - P(A \cap B)$. Note that the subtraction of $P(A \cap B)$ must be performed, or else it would result in double counting any outcomes that lie in both A and in B. For example, suppose that a 20-sided die is being rolled. Fred bets that the outcome will be greater than 10, while Helen bets that it will be greater than 4 but less than 15. What is the probability that at least one of them is correct?

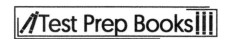

We apply the rule $P(A \cup B) = P(A) + P(B) - P(A \cap B)$, where A is that outcome x is in the range $x > 10$, and B is that outcome x is in the range $4 < x < 15$. $P(A) = 10 \times \frac{1}{20} = \frac{1}{2}$. $P(B) = 10 \times \frac{1}{20} = \frac{1}{2}$. $P(A \cap B)$ can be computed by noting that $A \cap B$ means the outcome x is in the range $10 < x < 15$, so $P(A \cap B) = 4 \times \frac{1}{20} = \frac{1}{5}$. Therefore, $P(A \cup B) = P(A) + P(B) - P(A \cap B) = \frac{1}{2} + \frac{1}{2} - \frac{1}{5} = \frac{4}{5}$.

Note that in this particular example, we could also have directly reasoned about the set of possible outcomes $A \cup B$, by noting that this would mean that x must be in the range $5 \leq x$. However, this is not always the case, depending on the given information.

Multiplication Rule

The *multiplication rule* for probabilities states the probability of A and B both happening is $P(A \cap B) = P(A)P(B|A)$. As an example, suppose that when Jamie wears black pants, there is a ½ probability that she wears a black shirt as well, and that she wears black pants ¾ of the time. What is the probability that she is wearing both a black shirt and black pants?

To figure this, use the above formula, where A will be "Jamie is wearing black pants," while B will be "Jamie is wearing a black shirt." It is known that $P(A)$ is ¾. It is also known that $P(B|A) = \frac{1}{2}$. Multiplying the two, the probability that she is wearing both black pants and a black shirt is $P(A)P(B|A) = \frac{3}{4} \times \frac{1}{2} = \frac{3}{8}$.

Reading

Literary Analysis

Style, Tone, and Mood

Style, tone, and mood are often thought to be the same thing. Though they're closely related, there are important differences to keep in mind. The easiest way to do this is to remember that style "creates and affects" tone and mood. More specifically, style is how the writer uses words to create the desired tone and mood for their writing.

Style

Style can include any number of technical writing choices. A few examples of style choices include:

- Sentence Construction: When presenting facts, does the writer use shorter sentences to create a quicker sense of the supporting evidence, or do they use longer sentences to elaborate and explain the information?

- Technical Language: Does the writer use jargon to demonstrate their expertise in the subject, or do they use ordinary language to help the reader understand things in simple terms?

- Formal Language: Does the writer refrain from using contractions such as *won't* or *can't* to create a more formal tone, or do they use a colloquial, conversational style to connect to the reader?

- Formatting: Does the writer use a series of shorter paragraphs to help the reader follow a line of argument, or do they use longer paragraphs to examine an issue in great detail and demonstrate their knowledge of the topic?

On the test, examine the writer's style and how their writing choices affect the way the text comes across.

Tone

Tone refers to the writer's attitude toward the subject matter. Tone is usually explained in terms of a work of fiction. For example, the tone conveys how the writer feels about their characters and the situations in which they're involved. Nonfiction writing is sometimes thought to have no tone at all; however, this is incorrect.

A lot of nonfiction writing has a neutral tone, which is an important tone for the writer to take. A neutral tone demonstrates that the writer is presenting a topic impartially and letting the information speak for itself. On the other hand, nonfiction writing can be just as effective and appropriate if the tone isn't neutral. For instance, take this example involving seat belts:

> Seat belts save more lives than any other automobile safety feature. Many studies show that airbags save lives as well; however, not all cars have airbags. For instance, some older cars don't. Furthermore, air bags aren't entirely reliable. For example, studies show that in 15% of accidents airbags don't deploy as designed, but, on the other hand, seat belt malfunctions are

extremely rare. The number of highway fatalities has plummeted since laws requiring seat belt usage were enacted.

In this passage, the writer mostly chooses to retain a neutral tone when presenting information. If the writer would instead include their own personal experience of losing a friend or family member in a car accident, the tone would change dramatically. The tone would no longer be neutral and would show that the writer has a personal stake in the content, allowing them to interpret the information in a different way. When analyzing tone, consider what the writer is trying to achieve in the text and how they *create* the tone using style.

Mood
Mood refers to the feelings and atmosphere that the writer's words create for the reader. Like tone, many nonfiction texts can have a neutral mood. To return to the previous example, if the writer would choose to include information about a person they know being killed in a car accident, the text would suddenly carry an emotional component that is absent in the previous example. Depending on how they present the information, the writer can create a sad, angry, or even hopeful mood. When analyzing the mood, consider what the writer wants to accomplish and whether the best choice was made to achieve that end.

Consistency

Whatever style, tone, and mood the writer uses, good writing should remain consistent throughout. If the writer chooses to include the tragic, personal experience above, it would affect the style, tone, and mood of the entire text. It would seem out of place for such an example to be used in the middle of a neutral, measured, and analytical text. To adjust the rest of the text, the writer needs to make additional choices to remain consistent. For example, the writer might decide to use the word *tragedy* in place of the more neutral *fatality*, or they could describe a series of car-related deaths as an *epidemic*. Adverbs and adjectives such as *devastating* or *horribly* could be included to maintain this consistent attitude toward the content. When analyzing writing, look for sudden shifts in style, tone, and mood, and consider whether the writer would be wiser to maintain the prevailing strategy.

Identify the Position and Purpose

When it comes to an author's writing, readers should always identify a position or stance. No matter how objective a text may seem, readers should assume the author has preconceived beliefs. One can reduce the likelihood of accepting an invalid argument by looking for multiple articles on the topic, including those with varying opinions. If several opinions point in the same direction and are backed by reputable peer-reviewed sources, it's more likely the author has a valid argument. Positions that run contrary to widely held beliefs and existing data should invite scrutiny. There are exceptions to the rule, so be a careful consumer of information.

Though themes, symbols, and motifs are buried deep within the text and can sometimes be difficult to infer, an author's purpose is usually obvious from the beginning. There are four purposes of writing: to inform, to persuade, to describe, and to entertain. Informative writing presents facts in an accessible way. Persuasive writing appeals to emotions and logic to inspire the reader to adopt a specific stance. Be wary of this type of writing, as it can mask a lack of objectivity with powerful emotion. Descriptive writing is designed to paint a picture in the reader's mind, while texts that entertain are often narratives designed to engage and delight the reader.

The various writing styles are usually blended, with one purpose dominating the rest. A persuasive text, for example, might begin with a humorous tale to make readers more receptive to the persuasive message, or a recipe in a cookbook designed to inform might be preceded by an entertaining anecdote that makes the recipes more appealing.

Identify Passage Characteristics

Writing can be classified under four passage types: narrative, expository, descriptive (sometimes called technical), and persuasive. Though these types are not mutually exclusive, one form tends to dominate the rest. By recognizing the *type* of passage you're reading, you gain insight into *how* you should read. When reading a narrative intended to entertain, sometimes you can read more quickly through the passage if the details are discernible. A technical document, on the other hand, might require a close read, because skimming the passage might cause the reader to miss salient details.

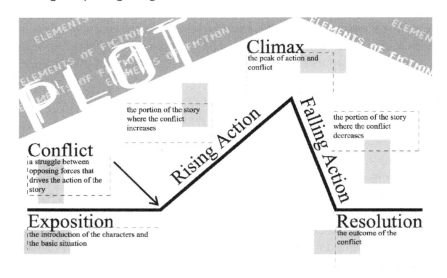

1. Narrative writing, at its core, is the art of storytelling. For a narrative to exist, certain elements must be present. First, it must have characters While many characters are human, characters could be defined as anything that thinks, acts, and talks like a human. For example, many recent movies, such as *Lord of the Rings* and *The Chronicles of Narnia*, include animals, fantasy creatures, and even trees that behave like humans. Narratives also must have a plot or sequence of events. Typically, those events follow a standard plot diagram, but recent trends start *in medias res* or in the middle (nearer the climax). In this instance, foreshadowing and flashbacks often fill in plot details. Finally, along with characters and a plot, there must also be conflict. Conflict is usually divided into two types: internal and external. Internal conflict indicates the character is in turmoil. Think of an angel on one shoulder and the devil on the other, arguing it out. Internal conflicts are presented through the character's thoughts. External conflicts are visible. Types of external conflict include person versus person, person versus nature, person versus technology, person versus the supernatural, or a person versus fate.

2. Expository writing is detached and to the point, while other types of writing — persuasive, narrative, and descriptive — are livelier. Since expository writing is designed to instruct or inform, it usually involves directions and steps written in second person ("you" voice) and lacks any persuasive or narrative elements. Sequence words such as *first*, *second*, and *third*, or *in the first place*, *secondly*, and

lastly are often given to add fluency and cohesion. Common examples of expository writing include instructor's lessons, cookbook recipes, and repair manuals.

3. Due to its empirical nature, technical writing is filled with steps, charts, graphs, data, and statistics. The goal of technical writing is to advance understanding in a field through the scientific method. Experts such as teachers, doctors, or mechanics use words unique to the profession in which they operate. These words, which often incorporate acronyms, are called *jargon*. Technical writing is a type of expository writing but is not meant to be understood by the general public. Instead, technical writers assume readers have received a formal education in a particular field of study and need no explanation as to what the jargon means. Imagine a doctor trying to understand a diagnostic reading for a car or a mechanic trying to interpret lab results. Only professionals with proper training will fully comprehend the text.

4. Persuasive writing is designed to change opinions and attitudes. The topic, stance, and arguments are found in the thesis, positioned near the end of the introduction. Later supporting paragraphs offer relevant quotations, paraphrases, and summaries from primary or secondary sources, which are then interpreted, analyzed, and evaluated. The goal of persuasive writers is not to stack quotes, but to develop original ideas by using sources as a starting point. Good persuasive writing makes powerful arguments with valid sources and thoughtful analysis. Poor persuasive writing is riddled with bias and logical fallacies. Sometimes logical and illogical arguments are sandwiched together in the same text. Therefore, readers should display skepticism when reading persuasive arguments.

Interpret Influences of Historical Context

Studying historical literature is fascinating. It reveals a snapshot in time of people, places, and cultures; a collective set of beliefs and attitudes that no longer exist. Writing changes as attitudes and cultures evolve. Beliefs previously considered immoral or wrong may be considered acceptable today. Researching the historical period of an author gives the reader perspective. The dialogue in Jane Austen's *Pride and Prejudice*, for example, is indicative of social class during the Regency era. Similarly, the stereotypes and slurs in *The Adventures of Huckleberry Finn* were a result of common attitudes and beliefs in the late 1800s, attitudes now found to be reprehensible.

Recognizing Cultural Themes

Regardless of culture, place, or time, certain themes are universal to the human condition. Because humans experience joy, rage, jealousy, and pride, certain themes span centuries. For example, Shakespeare's *Macbeth,* as well as modern works like *The 50ᵗʰ Law* by rapper 50 Cent and Robert Greene or the Netflix series *House of Cards* all feature characters who commit atrocious acts because of ambition. Similarly, *The Adventures of Huckleberry Finn*, published in the 1880s, and *The Catcher in the Rye*, published in the 1950s, both have characters who lie, connive, and survive on their wits.

Moviegoers know whether they are seeing an action, romance or horror film, and are often disappointed if the movie doesn't fit into the conventions of a particular category. Similarly, categories or genres give readers a sense of what to expect from a text. Some of the most basic genres in literature include books, short stories, poetry, and drama. Many genres can be split into sub-genres. For example, the sub-genres of historical fiction, realistic fiction, and fantasy all fit under the fiction genre.

Each genre has a unique way of approaching a particular theme. Books and short stories use plot, characterization, and setting, while poems rely on figurative language, sound devices, and symbolism. Dramas reveal plot through dialogue and the actor's voice and body language.

Main Ideas and Supporting Details

Topic Versus the Main Idea

It is very important to know the difference between the topic and the main idea of the text. Even though these two are similar because they both present the central point of a text, they have distinctive differences. A *topic* is the subject of the text; it can usually be described in a one- to two-word phrase and appears in the simplest form. On the other hand, the *main idea* is more detailed and provides the author's central point of the text. It can be expressed through a complete sentence and can be found in the beginning, middle, or end of a paragraph. In most nonfiction books, the first sentence of the passage usually (but not always) states the main idea. Take a look at the passage below to review the topic versus the main idea.

Cheetahs

Cheetahs are one of the fastest mammals on land, reaching up to 70 miles an hour over short distances. Even though cheetahs can run as fast as 70 miles an hour, they usually only have to run half that speed to catch up with their choice of prey. Cheetahs cannot maintain a fast pace over long periods of time because they will overheat their bodies. After a chase, cheetahs need to rest for approximately 30 minutes prior to eating or returning to any other activity.

In the example above, the topic of the passage is "Cheetahs" simply because that is the subject of the text. The main idea of the text is "Cheetahs are one of the fastest mammals on land but can only maintain this fast pace for short distances." While it covers the topic, it is more detailed and refers to the text in its entirety. The text continues to provide additional details called *supporting details,* which will be discussed in the next section.

Supporting Details

Supporting details help readers better develop and understand the main idea. Supporting details answer questions like *who, what, where, when, why,* and *how.* Different types of supporting details include examples, facts and statistics, anecdotes, and sensory details.

Persuasive and informative texts often use supporting details. In persuasive texts, authors attempt to make readers agree with their point of view, and supporting details are often used as "selling points." If authors make a statement, they should support the statement with evidence in order to adequately persuade readers. Informative texts use supporting details such as examples and facts to inform readers. Take another look at the previous "Cheetahs" passage to find examples of supporting details.

Cheetahs

Cheetahs are one of the fastest mammals on land, reaching up to 70 miles an hour over short distances. Even though cheetahs can run as fast as 70 miles an hour, they usually only have to run half that speed to catch up with their choice of prey. Cheetahs cannot maintain a fast pace over long periods of time because they will overheat their bodies. After a chase, cheetahs need to rest for approximately 30 minutes prior to eating or returning to any other activity.

In the example above, supporting details include:

- Cheetahs reach up to 70 miles per hour over short distances.
- They usually only have to run half that speed to catch up with their prey.
- Cheetahs will overheat their bodies if they exert a high speed over longer distances.
- Cheetahs need to rest for 30 minutes after a chase.

Look at the diagram below (applying the cheetah example) to help determine the hierarchy of topic, main idea, and supporting details.

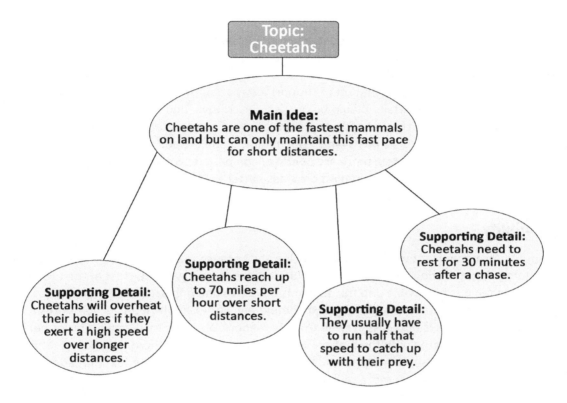

Drawing Conclusions

When drawing conclusions about texts or passages, readers should do two main things: 1) Use the information that they already know and 2) Use the information they have learned from the text or passage. Authors write with an intended purpose, and it is the reader's responsibility to understand and form logical conclusions of authors' ideas. It is important to remember that the reader's conclusions should be supported by information directly from the text. Readers cannot simply form conclusions based off of only information they already know.

There are several ways readers can draw conclusions from authors' ideas, such as note taking, text evidence, text credibility, writing a response to text, directly stated information versus implications, outlining, summarizing, and paraphrasing. Let's take a look at each important strategy to help readers draw logical conclusions.

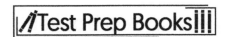

Note Taking

When readers take notes throughout texts or passages, they are jotting down important facts or points that the author makes. Note taking is a useful record of information that helps readers understand the text or passage and respond to it. When taking notes, readers should keep lines brief and filled with pertinent information so that they are not rereading a large amount of text, but rather just key points, elements, or words. After readers have completed a text or passage, they can refer to their notes to help them form a conclusion about the author's ideas in the text or passage.

Text Evidence

Text evidence is the information readers find in a text or passage that supports the main idea or point(s) in a story. In turn, text evidence can help readers draw conclusions about the text or passage. The information should be taken directly from the text or passage and placed in quotation marks. Text evidence provides readers with information to support ideas about the text so that they do not rely simply on their own thoughts. Details should be precise, descriptive, and factual. Statistics are a great piece of text evidence because they provide readers with exact numbers and not just a generalization. For example, instead of saying "Asia has a larger population than Europe," authors could provide detailed information such as, "In Asia there are over 4 billion people, whereas in Europe there are a little over 750 million." More definitive information provides better evidence to readers to help support their conclusions about texts or passages.

Text Credibility

Credible sources are important when drawing conclusions because readers need to be able to trust what they are reading. Authors should always use credible sources to help gain the trust of their readers. A text is *credible* when it is believable and the author is objective and unbiased. If readers do not trust an author's words, they may simply dismiss the text completely. For example, if an author writes a persuasive essay, he or she is outwardly trying to sway readers' opinions to align with his or her own. Readers may agree or disagree with the author, which may, in turn, lead them to believe that the author is credible or not credible. Also, readers should keep in mind the source of the text. If readers review a journal about astronomy, would a more reliable source be a NASA employee or a medical doctor? Overall, text credibility is important when drawing conclusions, because readers want reliable sources that support the decisions they have made about the author's ideas.

Writing a Response to Text

Once readers have determined their opinions and validated the credibility of a text, they can then reflect on the text. Writing a response to a text is one way readers can reflect on the given text or passage. When readers write responses to a text, it is important for them to rely on the evidence within the text to support their opinions or thoughts. Supporting evidence such as facts, details, statistics, and quotes directly from the text are key pieces of information readers should reflect upon or use when writing a response to text.

Directly Stated Information Versus Implications

Engaged readers should constantly self-question while reviewing texts to help them form conclusions. Self-questioning is when readers review a paragraph, page, passage, or chapter and ask themselves, "Did I understand what I read?," "What was the main event in this section?," "Where is this taking

52

place?," and so on. Authors can provide clues or pieces of evidence throughout a text or passage to guide readers toward a conclusion. This is why active and engaged readers should read the text or passage in its entirety before forming a definitive conclusion. If readers do not gather all the pieces of evidence needed, then they may jump to an illogical conclusion.

At times, authors directly state conclusions while others simply imply them. Of course, it is easier if authors outwardly provide conclusions to readers because it does not leave any information open to interpretation. On the other hand, implications are things that authors do not directly state but can be assumed based off of information they provided. If authors only imply what may have happened, readers can form a menagerie of ideas for conclusions. For example, look at the following statement: "Once we heard the sirens, we hunkered down in the storm shelter." In this statement, the author does not directly state that there was a tornado, but clues such as "sirens" and "storm shelter" provide insight to the readers to help form that conclusion.

Outlining

An outline is a system used to organize writing. When reading texts, outlining is important because it helps readers organize important information in a logical pattern using roman numerals. Usually, outlines start with the main idea(s) and then branch out into subgroups or subsidiary thoughts of subjects. Not only do outlines provide a visual tool for readers to reflect on how events, characters, settings, or other key parts of the text or passage relate to one another, but they can also lead readers to a stronger conclusion.

The sample below demonstrates what a general outline looks like.

I. Main Topic 1
   a. Subtopic 1
   b. Subtopic 2
      1. Detail 1
      2. Detail 2
II. Main Topic 2
   a. Subtopic 1
   b. Subtopic 2
      1. Detail 1
      2. Detail 2

Summarizing

At the end of a text or passage, it is important to summarize what the readers read. Summarizing is a strategy in which readers determine what is important throughout the text or passage, shorten those ideas, and rewrite or retell it in their own words. A summary should identify the main idea of the text or passage. Important details or supportive evidence should also be accurately reported in the summary. If writers provide irrelevant details in the summary, it may cloud the greater meaning of the summary in the text. When summarizing, writers should not include their opinions, quotes, or what they thought the author should have said. A clear summary provides clarity of the text or passage to the readers. Let's review the checklist of items writers should include in their summary.

Summary Checklist
- Title of the story
- Someone: Who is or are the main character(s)?
- Wanted: What did the character(s) want?
- But: What was the problem?
- So: How did the character(s) solve the problem?
- Then: How did the story end? What was the resolution?

Paraphrasing

Another strategy readers can use to help them fully comprehend a text or passage is paraphrasing. Paraphrasing is when readers take the author's words and put them into their own words. When readers and writers paraphrase, they should avoid copying the text—that is plagiarism. It is also important to include as many details as possible when restating the facts. Not only will this help readers and writers recall information, but by putting the information into their own words, they demonstrate whether or not they fully comprehend the text or passage. Look at the example below showing an original text and how to paraphrase it.

Original Text: Fenway Park is home to the beloved Boston Red Sox. The stadium opened on April 20, 1912. The stadium currently seats over 37,000 fans, many of whom travel from all over the country to experience the iconic team and nostalgia of Fenway Park.

Paraphrased: On April 20, 1912, Fenway Park opened. Home to the Boston Red Sox, the stadium now seats over 37,000 fans. Many spectators travel to watch the Red Sox and experience the spirit of Fenway Park.

Paraphrasing, summarizing, and quoting can often cross paths with one another. Review the chart below showing the similarities and differences between the three strategies.

Paraphrasing	Summarizing	Quoting
Uses own words	Puts main ideas into own words	Uses words that are identical to text
References original source	References original source	Requires quotation marks
Uses own sentences	Shows important ideas of source	Uses author's own words and ideas

Inferences in a Text

Readers should be able to make *inferences*. Making an inference requires the reader to read between the lines and look for what is *implied* rather than what is directly stated. That is, using information that is known from the text, the reader is able to make a logical assumption about information that is *not* directly stated but is probably true. Read the following passage:

"Hey, do you wanna meet my new puppy?" Jonathan asked.

"Oh, I'm sorry but please don't—" Jacinta began to protest, but before she could finish, Jonathan had already opened the passenger side door of his car and a perfect white ball of fur came bouncing towards Jacinta.

"Isn't he the cutest?" beamed Jonathan.

"Yes—achoo!—he's pretty—aaaachooo!!—adora—aaa—aaaachoo!" Jacinta managed to say in between sneezes. "But if you don't mind, I—I—achoo!—need to go inside."

Which of the following can be inferred from Jacinta's reaction to the puppy?
   a. she hates animals
   b. she is allergic to dogs
   c. she prefers cats to dogs
   d. she is angry at Jonathan

An inference requires the reader to consider the information presented and then form their own idea about what is probably true. Based on the details in the passage, what is the best answer to the question? Important details to pay attention to include the tone of Jacinta's dialogue, which is overall polite and apologetic, as well as her reaction itself, which is a long string of sneezes. Answer choices (a) and (d) both express strong emotions ("hates" and "angry") that are not evident in Jacinta's speech or actions. Answer choice (c) mentions cats, but there is nothing in the passage to indicate Jacinta's feelings about cats. Answer choice (b), "she is allergic to dogs," is the most logical choice—based on the fact that she began sneezing as soon as a fluffy dog approached her, it makes sense to guess that Jacinta might be allergic to dogs. So even though Jacinta never directly states, "Sorry, I'm allergic to dogs!" using the clues in the passage, it is still reasonable to guess that this is true.

Making inferences is crucial for readers of literature because literary texts often avoid presenting complete and direct information to readers about characters' thoughts or feelings, or they present this information in an unclear way, leaving it up to the reader to interpret clues given in the text. In order to make inferences while reading, readers should ask themselves:

- What details are being presented in the text?
- Is there any important information that seems to be missing?
- Based on the information that the author *does* include, what else is probably true?
- Is this inference reasonable based on what is already known?

Apply Information

A natural extension of being able to make an inference from a given set of information is also being able to apply that information to a new context. This is especially useful in non-fiction or informative writing. Considering the facts and details presented in the text, readers should consider how the same information might be relevant in a different situation. The following is an example of applying an inferential conclusion to a different context:

> Often, individuals behave differently in large groups than they do as individuals. One example of this is the psychological phenomenon known as the bystander effect. According to the bystander effect, the more people who witness an accident or crime occur, the less likely each individual bystander is to respond or offer assistance to the victim. A classic example of this is the murder of Kitty Genovese in New York City in the 1960s. Although there were over thirty witnesses to her killing by a stabber, none of them intervened to help Kitty or contact the police.

Considering the phenomenon of the bystander effect, what would probably happen if somebody tripped on the stairs in a crowded subway station?
a. Everybody would stop to help the person who tripped
b. Bystanders would point and laugh at the person who tripped
c. Someone would call the police after walking away from the station
d. Few if any bystanders would offer assistance to the person who tripped

This question asks readers to apply the information they learned from the passage, which is an informative paragraph about the bystander effect. According to the passage, this is a concept in psychology that describes the way people in groups respond to an accident—the more people are present, the less likely any one person is to intervene. While the passage illustrates this effect with the example of a woman's murder, the question asks readers to apply it to a different context—in this case, someone falling down the stairs in front of many subway passengers. Although this specific situation is not discussed in the passage, readers should be able to apply the general concepts described in the paragraph. The definition of the bystander effect includes any instance of an accident or crime in front of a large group of people. The question asks about a situation that falls within the same definition, so the general concept should still hold true: in the midst of a large crowd, few individuals are likely to actually respond to an accident. In this case, answer choice (d) is the best response.

Author's Use of Language

Authors utilize a wide range of techniques to tell a story or communicate information. Readers should be familiar with the most common of these techniques. Techniques of writing are also commonly known as rhetorical devices.

Types of Appeals

In non-fiction writing, authors employ argumentative techniques to present their opinion to readers in the most convincing way. First of all, persuasive writing usually includes at least one type of appeal: an appeal to logic (logos), emotion (pathos), or credibility and trustworthiness (ethos). When a writer appeals to logic, they are asking readers to agree with them based on research, evidence, and an established line of reasoning. An author's argument might also appeal to readers' emotions, perhaps by including personal stories and anecdotes (a short narrative of a specific event). A final type of appeal, appeal to authority, asks the reader to agree with the author's argument on the basis of their expertise or credentials. Consider three different approaches to arguing the same opinion:

Logic (Logos)
This is an example of an appeal to logic:

> Our school should abolish its current ban on cell phone use on campus. This rule was adopted last year as an attempt to reduce class disruptions and help students focus more on their lessons. However, since the rule was enacted, there has been no change in the number of disciplinary problems in class. Therefore, the rule is ineffective and should be done away with.

The author uses evidence to disprove the logic of the school's rule (the rule was supposed to reduce discipline problems; the number of problems has not been reduced; therefore, the rule is not working) and call for its repeal.

Emotion (Pathos)
An author's argument might also appeal to readers' emotions, perhaps by including personal stories and anecdotes. The next example presents an appeal to emotion. By sharing the personal anecdote of one student and speaking about emotional topics like family relationships, the author invokes the reader's empathy in asking them to reconsider the school rule.

> Our school should abolish its current ban on cell phone use on campus. If they aren't able to use their phones during the school day, many students feel isolated from their loved ones. For example, last semester, one student's grandmother had a heart attack in the morning. However, because he couldn't use his cell phone, the student didn't know about his grandmother's accident until the end of the day—when she had already passed away, and it was too late to say goodbye. By preventing students from contacting their friends and family, our school is placing undue stress and anxiety on students.

Credibility (Ethos)
Finally, an appeal to authority includes a statement from a relevant expert. In this case, the author uses a doctor in the field of education to support the argument. All three examples begin from the same opinion—the school's phone ban needs to change—but rely on different argumentative styles to persuade the reader.

> Our school should abolish its current ban on cell phone use on campus. According to Dr. Bartholomew Everett, a leading educational expert, "Research studies show that cell phone usage has no real impact on student attentiveness. Rather, phones provide a valuable technological resource for learning. Schools need to learn how to integrate this new technology into their curriculum." Rather than banning phones altogether, our school should follow the advice of experts and allow students to use phones as part of their learning.

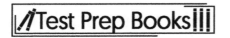

Rhetorical Questions

Another commonly used argumentative technique is asking rhetorical questions, questions that do not actually require an answer but that push the reader to consider the topic further.

> I wholly disagree with the proposal to ban restaurants from serving foods with high sugar and sodium contents. Do we really want to live in a world where the government can control what we eat? I prefer to make my own food choices.

Here, the author's rhetorical question prompts readers to put themselves in a hypothetical situation and imagine how they would feel about it.

Figurative Language

Literary texts also employ rhetorical devices. Figurative language like simile and metaphor is a type of rhetorical device commonly found in literature. In addition to rhetorical devices that play on the *meanings* of words, there are also rhetorical devices that use the *sounds* of words. These devices are most often found in poetry but may also be found in other types of literature and in non-fiction writing like speech texts.

Alliteration and *assonance* are both varieties of sound repetition. Other types of sound repetition include: anaphora, repetition that occurs at the beginning of the sentences; epiphora, repetition occurring at the end of phrases; antimetabole, repetition of words in reverse order; and antiphrasis, a form of denial of an assertion in a text.

Alliteration refers to the repetition of the first sound of each word. Recall Robert Burns' opening line:

> My love is like a red, red rose

This line includes two instances of alliteration: "love" and "like" (repeated *L* sound), as well as "red" and "rose" (repeated *R* sound). Next, assonance refers to the repetition of vowel sounds, and can occur anywhere within a word (not just the opening sound). Here is the opening of a poem by John Keats:

> When I have fears that I may cease to be

> Before my pen has glean'd my teeming brain

Assonance can be found in the words "fears," "cease," "be," "glean'd," and "teeming," all of which stress the long *E* sound. Both alliteration and assonance create a harmony that unifies the writer's language.

Another sound device is *onomatopoeia*, or words whose spelling mimics the sound they describe. Words such as "crash," "bang," and "sizzle" are all examples of onomatopoeia. Use of onomatopoetic language adds auditory imagery to the text.

Readers are probably most familiar with the technique of *pun*. A pun is a play on words, taking advantage of two words that have the same or similar pronunciation. Puns can be found throughout Shakespeare's plays, for instance:

> Now is the winter of our discontent
> Made glorious summer by this son of York

These lines from *Richard III* contain a play on words. Richard III refers to his brother, the newly crowned King Edward IV, as the "son of York," referencing their family heritage from the house of York. However, while drawing a comparison between the political climate and the weather (times of political trouble were the "winter," but now the new king brings "glorious summer"), Richard's use of the word "son" also implies another word with the same pronunciation, "sun"—so Edward IV is also like the sun, bringing light, warmth, and hope to England. Puns are a clever way for writers to suggest two meanings at once.

Counterarguments

If an author presents a differing opinion or a counterargument in order to refute it, the reader should consider how and why this information is being presented. It is meant to strengthen the original argument and shouldn't be confused with the author's intended conclusion, but it should also be considered in the reader's final evaluation.

Authors can also use bias if they ignore the opposing viewpoint or present their side in an unbalanced way. A strong argument considers the opposition and finds a way to refute it. Critical readers should look for an unfair or one-sided presentation of the argument and be skeptical, as a bias may be present. Even if this bias is unintentional, if it exists in the writing, the reader should be wary of the validity of the argument. Readers should also look for the use of stereotypes, which refer to specific groups. Stereotypes are often negative connotations about a person or place, and should always be avoided. When a critical reader finds stereotypes in a piece of writing, they should be critical of the argument, and consider the validity of anything the author presents. Stereotypes reveal a flaw in the writer's thinking and may suggest a lack of knowledge or understanding about the subject.

Meaning of Words in Context

There will be many occasions in one's reading career in which an unknown word or a word with multiple meanings will pop up. There are ways of determining what these words or phrases mean that do not require the use of the dictionary, which is especially helpful during a test where one may not be available. Even outside of the exam, knowing how to derive an understanding of a word via context clues will be a critical skill in the real world. The context is the circumstances in which a story or a passage is happening, and can usually be found in the series of words directly before or directly after the word or phrase in question. The clues are the words that hint towards the meaning of the unknown word or phrase.

There may be questions that ask about the meaning of a particular word or phrase within a passage. There are a couple ways to approach these kinds of questions:

1. Define the word or phrase in a way that is easy to comprehend (using context clues).
2. Try out each answer choice in place of the word.

To demonstrate, here's an example from *Alice in Wonderland*:

> Alice was beginning to get very tired of sitting by her sister on the bank, and of having nothing to do: once or twice she <u>peeped</u> into the book her sister was reading, but it had no pictures or conversations in it, "and what is the use of a book," thought Alice, "without pictures or conversations?"

Q: As it is used in the selection, the word <u>peeped</u> means:

Using the first technique, before looking at the answers, define the word "peeped" using context clues and then find the matching answer. Then, analyze the entire passage in order to determine the meaning, not just the surrounding words.

To begin, imagine a blank where the word should be and put a synonym or definition there: "once or twice she _____ into the book her sister was reading." The context clue here is the book. It may be tempting to put "read" where the blank is, but notice the preposition word, "into." One does not read *into* a book, one simply reads a book, and since reading a book requires that it is seen with a pair of eyes, then "look" would make the most sense to put into the blank: "once or twice she <u>looked </u>into the book her sister was reading."

Once an easy-to-understand word or synonym has been supplanted, readers should check to make sure it makes sense with the rest of the passage. What happened after she looked into the book? She thought to herself how a book without pictures or conversations is useless. This situation in its entirety makes sense.

Now check the answer choices for a match:
- a. To make a high-pitched cry
- b. To smack
- c. To look curiously
- d. To pout

Since the word was already defined, Choice *C* is the best option.

Using the second technique, replace the figurative blank with each of the answer choices and determine which one is the most appropriate. Remember to look further into the passage to clarify that they work, because they could still make sense out of context.
- a. Once or twice she <u>made a high pitched cry</u> into the book her sister was reading
- b. Once or twice she <u>smacked</u> into the book her sister was reading
- c. Once or twice she <u>looked curiously</u> into the book her sister was reading
- d. Once or twice she <u>pouted</u> into the book her sister was reading

For Choice *A*, it does not make much sense in any context for a person to yell into a book, unless maybe something terrible has happened in the story. Given that afterward Alice thinks to herself how useless a book without pictures is, this option does not make sense within context.

For Choice *B*, smacking a book someone is reading may make sense if the rest of the passage indicates a reason for doing so. If Alice was angry or her sister had shoved it in her face, then maybe smacking the book would make sense within context. However, since whatever she does with the book causes her to think, "what is the use of a book without pictures or conversations?" then answer Choice *B* is not an appropriate answer. Answer Choice *C* fits well within context, given her subsequent thoughts on the matter. Answer Choice *D* does not make sense in context or grammatically, as people do not "pout into" things.

This is a simple example to illustrate the techniques outlined above. There may, however, be a question in which all of the definitions are correct and also make sense out of context, in which the appropriate context clues will really need to be honed in on in order to determine the correct answer. For example, here is another passage from *Alice in Wonderland*:

> . . . but when the Rabbit actually took a watch out of its waistcoat pocket, and looked at it, and then hurried on, Alice <u>started</u> to her feet, for it flashed across her mind that she had never before seen a rabbit with either a waistcoat-pocket or a watch to take out of it, and burning with curiosity, she ran across the field after it, and was just in time to see it pop down a large rabbit-hole under the hedge.

Q: As it is used in the passage, the word started means
    a. To turn on
    b. To begin
    c. To move quickly
    d. To be surprised

All of these words qualify as a definition of "start," but using context clues, the correct answer can be identified using one of the two techniques above. It's easy to see that one does not turn on, begin, or be surprised to one's feet. The selection also states that she "ran across the field after it," indicating that she was in a hurry. Therefore, to move quickly would make the most sense in this context.

The same strategies can be applied to vocabulary that may be completely unfamiliar. In this case, focus on the words before or after the unknown word in order to determine its definition. Take this sentence, for example:

> Sam was such a <u>miser</u> that he forced Andrew to pay him twelve cents for the candy, even though he had a large inheritance and he knew his friend was poor.

Unlike with assertion questions, for vocabulary questions, it may be necessary to apply some critical thinking skills that may not be explicitly stated within the passage. Think about the implications of the passage, or what the text is trying to say. With this example, it is important to realize that it is considered unusually stingy for a person to demand so little money from someone instead of just letting their friend have the candy, especially if this person is already wealthy. Hence, a <u>miser</u> is a greedy or stingy individual.

Questions about complex vocabulary may not be explicitly asked, but this is a useful skill to know. If there is an unfamiliar word while reading a passage and its definition goes unknown, it is possible to miss out on a critical message that could inhibit the ability to appropriately answer the questions. Practicing this technique in daily life will sharpen this ability to derive meanings from context clues with ease.

Writing

Essay Revision

Revisions

Leaving a few minutes at the end to revise and proofread offers an opportunity for writers to polish things up. Putting one's self in the reader's shoes and focusing on what the essay actually says helps writers identify problems—it's a movement from the mindset of writer to the mindset of editor. The goal is to have a clean, clear copy of the essay. The following areas should be considered when proofreading:

- Sentence fragments
- Awkward sentence structure
- Run-on sentences
- Incorrect word choice
- Grammatical agreement errors
- Spelling errors
- Punctuation errors
- Capitalization errors

Organization

Good writing is not merely a random collection of sentences. No matter how well written, sentences must relate and coordinate appropriately with one another. If not, the writing seems random, haphazard, and disorganized. Therefore, good writing must be organized, where each sentence fits a larger context and relates to the sentences around it.

Transition Words

The writer should act as a guide, showing the reader how all the sentences fit together. Consider the seat belt example again:

> Seat belts save more lives than any other automobile safety feature. Many studies show that airbags save lives as well. Not all cars have airbags. Many older cars don't. Air bags aren't entirely reliable. Studies show that in 15% of accidents, airbags don't deploy as designed. Seat belt malfunctions are extremely rare.

There's nothing wrong with any of these sentences individually, but together they're disjointed and difficult to follow. The best way for the writer to communicate information is through the use of transition words. Here are examples of transition words and phrases that tie sentences together, enabling a more natural flow:

- To show causality: *as a result, therefore*, and *consequently*
- To compare and contrast: *however, but*, and *on the other hand*
- To introduce examples: *for instance, namely*, and *including*
- To show order of importance: *foremost, primarily, secondly*, and *lastly*

NOTE: This is not a complete list of transitions. There are many more that can be used; however, most fit into these or similar categories. The important point is that the words should clearly show the relationship between sentences, supporting information, and the main idea.

Here is an update to the previous example using transition words. These changes make it easier to read and bring clarity to the writer's points:

> Seat belts save more lives than any other automobile safety feature. Many studies show that airbags save lives as well; however, not all cars have airbags. For instance, some older cars don't. Furthermore, air bags aren't entirely reliable. For example, studies show that in 15% of accidents, airbags don't deploy as designed, but, on the other hand, seat belt malfunctions are extremely rare.

Also be prepared to analyze whether the writer is using the best transition word or phrase for the situation. Take this sentence for example: "As a result, seat belt malfunctions are extremely rare." This sentence doesn't make sense in the context above because the writer is trying to show the contrast between seat belts and airbags, not the causality.

Logical Sequence

Even if the writer includes plenty of information to support their point, the writing is only coherent when the information is in a logical order. First, the writer should introduce the main idea, whether for a paragraph, a section, or the entire piece. Second, they should present evidence to support the main idea by using transitional language. This shows the reader how the information relates to the main idea and to the sentences around it. The writer should then take time to interpret the information, making sure necessary connections are obvious to the reader. Finally, the writer can summarize the information in a closing section.

Though most writing follows this pattern, it isn't a set rule. Sometimes writers change the order for effect. For example, the writer can begin with a surprising piece of supporting information to grab the reader's attention, and then transition to the main idea. Thus, if a passage doesn't follow the logical order, don't immediately assume it's wrong. However, most writing usually settles into a logical sequence after a nontraditional beginning.

Introductions and Conclusions

Examining the writer's strategies for introductions and conclusions puts the reader in the right mindset to interpret the rest of the text. Look for methods the writer might use for introductions such as:

- Stating the main point immediately, followed by outlining how the rest of the piece supports this claim.

- Establishing important, smaller pieces of the main idea first, and then grouping these points into a case for the main idea.

- Opening with a quotation, anecdote, question, seeming paradox, or other piece of interesting information, and then using it to lead to the main point.

Whatever method the writer chooses, the introduction should make their intention clear, establish their voice as a credible one, and encourage a person to continue reading.

Conclusions tend to follow a similar pattern. In them, the writer restates their main idea a final time, often after summarizing the smaller pieces of that idea. If the introduction uses a quote or anecdote to grab the reader's attention, the conclusion often makes reference to it again. Whatever way the writer chooses to arrange the conclusion, the final restatement of the main idea should be clear and simple for the reader to interpret. Finally, conclusions shouldn't introduce any new information.

Precision

People often think of precision in terms of math, but precise word choice is another key to successful writing. Since language itself is imprecise, it's important for the writer to find the exact word or words to convey the full, intended meaning of a given situation. For example:

> The number of deaths has gone down since seat belt laws started.

There are several problems with this sentence. First, the word *deaths* is too general. From the context, it's assumed that the writer is referring only to deaths caused by car accidents. However, without clarification, the sentence lacks impact and is probably untrue. The phrase "gone down" might be accurate, but a more precise word could provide more information and greater accuracy. Did the numbers show a slow and steady decrease of highway fatalities or a sudden drop? If the latter is true, the writer is missing a chance to make their point more dramatically. Instead of "gone down" they could substitute *plummeted*, *fallen drastically*, or *rapidly diminished* to bring the information to life. Also, the phrase "seat belt laws" is unclear. Does it refer to laws requiring cars to include seat belts or to laws requiring drivers and passengers to use them? Finally, *started* is not a strong verb. Words like *enacted* or *adopted* are more direct and make the content more real. When put together, these changes create a far more powerful sentence:

> The number of highway fatalities has plummeted since laws requiring seat belt usage were enacted.

However, it's important to note that precise word choice can sometimes be taken too far. If the writer of the sentence above takes precision to an extreme, it might result in the following:

The incidence of high-speed, automobile accident related fatalities has decreased 75% and continued to remain at historical lows since the initial set of federal legislations requiring seat belt use were enacted in 1992.

This sentence is extremely precise, but it takes so long to achieve that precision that it suffers from a lack of clarity. Precise writing is about finding the right balance between information and flow. This is also an issue of conciseness (discussed in the next section).

The last thing to consider with precision is a word choice that's not only unclear or uninteresting, but also confusing or misleading. For example:

The number of highway fatalities has become hugely lower since laws requiring seat belt use were enacted.

In this case, the reader might be confused by the word *hugely*. Huge means large, but here the writer uses *hugely* to describe something small. Though most readers can decipher this, doing so disconnects them from the flow of the writing and makes the writer's point less effective.

Conciseness

"Less is more" is a good rule to follow when writing a sentence. Unfortunately, writers often include extra words and phrases that seem necessary at the time, but add nothing to the main idea. This confuses the reader and creates unnecessary repetition. Writing that lacks conciseness is usually guilty of excessive wordiness and redundant phrases. Here's an example containing both of these issues:

> When legislators decided to begin creating legislation making it mandatory for automobile drivers and passengers to make use of seat belts while in cars, a large number of them made those laws for reasons that were political reasons.

There are several empty or "fluff" words here that take up too much space. These can be eliminated while still maintaining the writer's meaning. For example:

- "Decided to begin" could be shortened to "began"
- "Making it mandatory for" could be shortened to "requiring"
- "Make use of" could be shortened to "use"
- "A large number" could be shortened to "many"

In addition, there are several examples of redundancy that can be eliminated:

- "Legislators decided to begin creating legislation" and "made those laws"
- "Automobile drivers and passengers" and "while in cars"
- "Reasons that were political reasons"

These changes are incorporated as follows:

> When legislators began requiring drivers and passengers to use seat belts, many of them did so for political reasons.

There are many general examples of redundant phrases, such as "add an additional," "complete and total," "time schedule," and "transportation vehicle." If asked to identify a redundant phrase on the test, look for words that are close together with the same (or similar) meanings.

Agreement

Agreement in Number

Subjects and verbs must agree in number. If a sentence has a singular subject, then it must use a singular verb. If there is a plural subject, then it must use a plural verb.

Singular Noun	Singular Verb	Plural Noun	Plural verb
Man	has	men	have
child	plays	children	play
basketball	bounces	basketballs	bounce

Agreement in Person

Verbs must also agree in person. A subject that uses first person must include a verb that is also in first person. The same is true for second and third person subjects and verbs.

	Noun	Verb
First person	I	am
Second person	you	are
Third person	she	is

Common Agreement Errors

Compound Subjects

Compound subjects are when two or more subjects are joined by a coordinating conjunction, such as *and, or, neither,* or *nor.* Errors in agreement sometimes occur when there is a compound subject:

Incorrect: Mike and I am in a meeting this morning.

Correct: Mike and I are in a meeting this morning.

A compound subject always uses the plural form of the verb to match with the plural subject. In the above example, readers can substitute "Mike and I" for "we" to make it easier to determine the verb: "We *are* in a meeting this morning."

Separation of Subject and Verb

Errors sometimes occur when the subject is separated from the verb by a prepositional phrase or parenthetical element:

Incorrect: The objective of the teachers are to help students learn.

Correct: The objective of the teachers is to help students learn.

The verb must agree with the singular subject *objective*, not the word *teachers*, which is the object of the preposition *of* and does not influence the subject. An easy way to determine if the subject and verb agree is to take out the middle preposition: "The objective *is* to help students learn."

Indefinite Pronouns

Indefinite pronouns refer to people or groups in a general way: *each, anyone, none, all, either, neither,* and *everyone.* Some indefinite pronouns are always singular, such as *each, everyone, someone,* and *everybody,* which affects verb choice:

Incorrect: Each of them are competing in the race.

Correct: Each of them is competing in the race.

While the word *them* can indicate that a plural verb is needed, the subject *each* is singular regardless of what it refers to, requiring the singular verb, *is.*

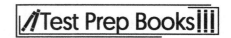

Other indefinite pronouns can be singular or plural, depending on what they are referring to, such as *anyone, all,* and *some.*

Some of the orders are scheduled to arrive today.

Some refers to *orders*, which is plural, so the plural verb (*are*) is needed.

Some of the cake is left on the dining room table.

Some refers to *cake*, which is singular, so the singular verb (*is*) is needed.

Subjects Joined by Or and Nor
Compound subjects joined by *or* or *nor* rely on the subject nearest to the verb to determine conjugation and agreement:

Neither Ben nor Jeff was in attendance at the conference.

Pink or purple is the bride's color choice.

In each example, the subjects are both singular, so the verb should be singular.

If one subject is singular and the other plural, the subject nearest to the verb is the one that needs to agree:

Either the shirt or pants are hanging on the clothesline.

In this example there is a singular subject (*shirt*) and a plural subject (*pants*), so the verb (*are*) should agree with the subject nearest to it (*pants*).

Collective Nouns
Collective nouns can use a singular or plural verb depending on their function in the sentence. If the collective noun is acting as a unit, then a singular verb is needed. Otherwise, it's necessary to use a plural verb.

The staff is required to meet every third Friday of the month.

The *staff* is meeting as a collective unit, so a singular verb is needed.

The staff are getting in their cars to go home.

The staff get into their cars separately, so a plural verb is needed.

Plural Nouns with Singular Meaning
Certain nouns end in *s*, like a plural noun, but have singular meaning, such as *mathematics, news,* and *civics*. These nouns should use a singular verb.

The news is on at 8:00 tonight.

Nouns that are single things, but have two parts, are considered plural and should use a plural verb, such as *scissors, pants,* and *tweezers.*

My favorite pants are in the washing machine.

There Is and There Are

There cannot be a subject, so verb agreement should be based on a word that comes after the verb.

There is a hole in the road.

The subject in this sentence is *hole*, which is singular, so the verb should be singular (*is*).

There are kids playing kickball in the street.

The subject in this sentence is *kids*, which is plural, so the verb should be plural (*are*).

Verb Tense

Shifting verb forms entails conjugation, which is used to indicate tense, voice, or mood.

Verb tense is used to show when the action in the sentence took place. There are several different verb tenses, and it is important to know how and when to use them. Some verb tenses can be achieved by changing the form of the verb, while others require the use of helping verbs (e.g., *is, was,* or *has*).

- *Present tense* shows the action is happening currently or is ongoing:

  I walk to work every morning.

  She is stressed about the deadline.

- *Past tense* shows that the action happened in the past or that the state of being is in the past:

  I walked to work yesterday morning.

  She was stressed about the deadline.

- *Future tense* shows that the action will happen in the future or is a future state of being:

  I will walk to work tomorrow morning.

  She will be stressed about the deadline.

- *Present perfect tense* shows action that began in the past, but continues into the present:

  I have walked to work all week.

  She has been stressed about the deadline.

- *Past perfect tense* shows an action was finished before another took place:

  I had walked all week until I sprained my ankle.

  She had been stressed about the deadline until we talked about it.

- *Future perfect tense* shows an action that will be completed at some point in the future:

  By the time the bus arrives, I will have walked to work already.

Sentence Structure

Sentence Types

There are four ways in which we can structure sentences: simple, compound, complex, and compound-complex. Sentences can be composed of just one clause or many clauses joined together.

When a sentence is composed of just one clause (an independent clause), we call it a simple sentence. Simple sentences do not necessarily have to be short sentences. They just require one independent clause with a subject and a predicate. For example:

Thomas marched over to Andrew's house.

Jonah and Mary constructed a simplified version of the Eiffel Tower with Legos.

When a sentence has two or more independent clauses we call it a compound sentence. The clauses are connected by a comma and a coordinating conjunction—*and, but, or, nor, for*—or by a semicolon. Compound sentences do not have dependent clauses. For example:

We went to the fireworks stand, and we bought enough fireworks to last all night.

The children sat on the grass, and then we lit the fireworks one at a time.

When a sentence has just one independent clause and includes one or more dependent clauses, we call it a complex sentence:

Because she slept well and drank coffee, Sarah was quite productive at work.

Although Will had coffee, he made mistakes while using the photocopier.

When a sentence has two or more independent clauses and at least one dependent clause, we call it a compound-complex sentence:

It may come as a surprise, but I found the tickets, and you can go to the show.

Jade is the girl who dove from the high-dive, and she stunned the audience silent.

Sentence Fragments

Remember that a complete sentence must have both a subject and a verb. Complete sentences consist of at least one independent clause. Incomplete sentences are called sentence fragments. A sentence fragment is a common error in writing. Sentence fragments can be independent clauses that start with subordinating words, such as *but, as, so that,* or *because,* or they could simply be missing a subject or verb.

You can correct a fragment error by adding the fragment to a nearby sentence or by adding or removing words to make it an independent clause. For example:

Dogs are my favorite animals. Because cats are too independent. (Incorrect; the word because creates a sentence fragment)

Dogs are my favorite animals because cats are too independent. (Correct; the fragment becomes a dependent clause.)

Dogs are my favorite animals. Cats are too independent. (Correct; the fragment becomes a simple sentence.)

Run-on Sentences

Another common mistake in writing is the run-on sentence. A run-on is created when two or more independent clauses are joined without the use of a conjunction, a semicolon, a colon, or a dash. We don't want to use commas where periods belong. Here is an example of a run-on sentence:

Making wedding cakes can take many hours I am very impatient, I want to see them completed right away.

There are a variety of ways to correct a run-on sentence. The method you choose will depend on the context of the sentence and how it fits with neighboring sentences:

Making wedding cakes can take many hours. I am very impatient. I want to see them completed right away. (Use periods to create more than one sentence.)

Making wedding cakes can take many hours; I am very impatient—I want to see them completed right away. (Correct the sentence using a semicolon, colon, or dash.)

Making wedding cakes can take many hours and I am very impatient, so I want to see them completed right away. (Correct the sentence using coordinating conjunctions.)

I am very impatient because I would rather see completed wedding cakes right away than wait for it to take many hours. (Correct the sentence by revising.)

Dangling and Misplaced Modifiers

A modifier is a word or phrase meant to describe or clarify another word in the sentence. When a sentence has a modifier but is missing the word it describes or clarifies, it's an error called a dangling modifier. We can fix the sentence by revising to include the word that is being modified. Consider the following examples with the modifier italicized:

Having walked five miles, this bench will be the place to rest. (Incorrect; this version of the sentence implies that the bench walked the miles, not the person.)

Having walked five miles, Matt will rest on this bench. (Correct; in this version, *having walked five miles* correctly modifies *Matt*, who did the walking.)

Since midnight, my dreams have been pleasant and comforting. (Incorrect; in this version, the adverb clause *since midnight* cannot modify the noun *dreams.*)

Since midnight, I have had pleasant and comforting dreams. (Correct; in this version, *since midnight* modifies the verb *have had*, telling us when the dreams occurred.)

Sometimes the modifier is not located close enough to the word it modifies for the sentence to be clearly understood. In this case, we call the error a misplaced modifier. Here is an example with the modifier italicized and the modified word in underlined.

We gave the hot <u>cocoa</u> to the children *that was filled with marshmallows.* (Incorrect; this sentence implies that the children are what are filled with marshmallows.)

We gave the hot <u>*cocoa* *that was filled with marshmallows*</u> to the children. (Correct; here, the cocoa is filled with marshmallows. The modifier is near the word it modifies.)

Parallelism and Subordination

Parallelism

To be grammatically correct we must use articles, prepositions, infinitives, and introductory words for dependent clauses consistently throughout a sentence. This is called parallelism. We use parallelism when we are matching parts of speech, phrases, or clauses with another part of the sentence. Being inconsistent creates confusion. Consider the following example.

Incorrect: Be ready for running and to ride a bike during the triathlon.

Correct: Be ready to run and to ride a bike during the triathlon.

Correct: Be ready for running and for riding a bike during the triathlon.

In the incorrect example, the gerund *running* does not match with the infinitive *to ride.* Either both should be infinitives or both should be gerunds.

Subordination

Sometimes we have unequal pieces of information in a sentence where one piece is more important than the other. We need to show that one piece of information is subordinate to the other. We can make the more important piece an independent clause and connect the other piece by making it a dependent clause. Consider this example:

Central thought: Kittens can always find their mother.

Subordinate: Kittens are blind at birth.

Complex Sentence: Despite being blind at birth, kittens can always find their mother.

The sentence "Kittens are blind at birth" is made subordinate to the sentence "Kittens can always find their mother" by placing the word "Despite" at the beginning and removing the subject, thus turning an independent clause ("kittens are blind at birth") into a subordinate phrase ("Despite being blind at birth").

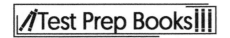

Sentence Logic

Clauses

Clauses are groups of words within a sentence that have both a subject and a verb. We can distinguish a clause from a phrase because phrases do not have both a subject and a verb. There are several types of clauses; clauses can be independent or dependent and can serve as a noun, an adjective, or an adverb.

An *independent clause* could stand alone as its own sentence if the rest of the sentence were not there. For example:

> *The party is on Tuesday* after the volleyball game is over.

> *I am excited to go to the party* because my best friend will be there.

A *dependent clause*, or subordinating clause, is the part of the sentence that gives supportive information but cannot create a proper sentence by itself. However, it will still have both a subject and a verb; otherwise, it is a phrase. In the example above, *after the volleyball game is over* and *because my best friend will be there* are dependent because they begin with the conjunctions *after* and *because*, and a proper sentence does not begin with a conjunction.

Noun clauses are groups of words that collectively form a noun. Look for the opening words *whether, which, what, that, who, how,* or *why.* For example:

> I had fun cooking *what we had for dinner last night.*

> I'm going to track down *whoever ate my sandwich.*

Adjective clauses collectively form an adjective that modifies a noun or pronoun in the sentence. If you can remove the adjective clause and the leftovers create a standalone sentence, then the clause should be set off with commas, parentheses, or dashes. If you can remove the clause it is called nonrestrictive. If it can't be removed without ruining the sentence, then it is called restrictive and does not get set off with commas.

> Jenna, *who hates to get wet,* fell into the pool. (Nonrestrictive)

> The girl *who hates to get wet* fell into the pool. (Restrictive; the clause tells us which girl, and if removed there is confusion)

Adverbial clauses serve as an adverb in the sentence, modifying a verb, adjective, or other adverb. Look for the opening words *after, before, as, as if, although, because, if, since, so, so that, when, where, while,* or *unless.*

> She lost her wallet after she left the theme park.

> Her earring fell through the crack before she could catch it.

Phrases

A phrase is a group of words that go together but do not include both a subject and a verb. We use them to add information, explain something, or make the sentence easier for the reader to understand. Unlike clauses, phrases cannot ever stand alone as their own sentence if the rest of the sentence were not there. They do not form complete thoughts. There are noun phrases, prepositional phrases, verbal phrases, appositive phrases, and absolute phrases. Let's look at each of these.

Noun phrases: A noun phrase is a group of words built around a noun or pronoun that serves as a unit to form a noun in the sentence. Consider the following examples. The phrase is built around the underlined word. The entire phrase can be replaced by a noun or pronoun to test whether or not it is a noun phrase.

> I like the chocolate chip ice cream. (I like it.)

> I know all the shortest routes. (I know them.)

> I met the best supporting actress. (I met her.)

Prepositional phrases: These are phrases that begin with a preposition and end with a noun or pronoun. We use them as a unit to form the adjective or adverb in the sentence. Prepositional phrases that introduce a sentence are called introductory prepositional phrases and are set off with commas.

> I found the Frisbee *on the roof peak.* (Adverb; where it was found)

> The girl *with the bright red hair* was prom queen. (Adjective; which girl)

> *Before the sequel,* we wanted to watch the first movie. (Introductory phrase)

Verbal phrases: Some phrases look like verbs but do not serve as the verb in the sentence. These are called verbal phrases. There are three types: participial phrases, gerund phrases, and infinitive phrases.

Participial phrases start with a participle and modify nouns or pronouns; therefore, they act as the adjective in the sentence.

> *Beaten by the sun,* we searched for shade to sit in. (Modifies the pronoun *we*)

> The hikers, *being eaten by mosquitoes,* longed for repellant. (Modifies the noun *hikers*)

Gerund phrases often look like participles because they end in *-ing*, but they serve as the noun, not the adjective, in the sentence. Like any noun, we can use them as the subject or as the object of a verb or preposition in the sentence.

> *Eating green salad* is the best way to lose weight. (Subject)

> Sumo wrestlers are famous for *eating large quantities of food.* (Object)

Infinitive phrases often look like verbs because they start with the word *to,* but they serve as an adjective, adverb, or noun.

> *To survive the chill* is the goal of the Polar Bear Plunge. (Noun)

> A hot tub is at the scene *to warm up after the jump.* (Adverb)

> The jumpers have hot cocoa *to drink right away.* (Adjective)

Appositive phrases: We can use any of the above types of phrases to rename nouns or pronouns, and we call this an appositive phrase. Appositive phrases usually appear either just before or just after the noun or pronoun they are renaming. Appositive phrases are essential when the noun or pronoun is too general, and they are nonessential when they just add information.

> *The two famous brothers* Orville and Wilbur Wright invented the airplane. (Essential)

> Sarah Calysta, *my great grandmother,* is my namesake. (Nonessential)

Absolute phrases: When a participle comes after a noun and forms a phrase that is not otherwise part of the sentence, it's called an absolute phrase. Absolute phrases are not complete thoughts and cannot stand alone because they do not have a subject and a verb. They are not essential to the sentence in that they do not explain or add additional meaning to any other part of the sentence.

> *The engine roaring,* Jada closed her eyes and waited for the plane to take off.

> *The microphone crackling,* the flight attendant announced the delayed arrival.

Five-Paragraph Persuasive Essay

Brainstorming

One of the most important steps in writing an essay is prewriting. Before drafting an essay, it's helpful to think about the topic for a moment or two, in order to gain a more solid understanding of what the task is. Then, spending about five minutes jotting down the immediate ideas that could work for the essay is recommended. It is a way to get some words on the page and offer a reference for ideas when drafting. Scratch paper is provided for writers to use any prewriting techniques such as webbing, free writing, or listing. The goal is to get ideas out of the mind and onto the page.

Considering Opposing Viewpoints

In the planning stage, it's important to consider all aspects of the topic, including different viewpoints on the subject. There are more than two ways to look at a topic, and a strong argument considers those opposing viewpoints. Considering opposing viewpoints can help writers present a fair, balanced, and informed essay that shows consideration for all readers. This approach can also strengthen an argument by recognizing and potentially refuting the opposing viewpoint(s).

Drawing from personal experience may help to support ideas. For example, if the goal for writing is a personal narrative, then the story should be from the writer's own life. Many writers find it helpful to draw from personal experience, even in an essay that is not strictly narrative. Personal anecdotes or short stories can help to illustrate a point in other types of essays as well.

Moving from Brainstorming to Planning

Once the ideas are on the page, it's time to turn them into a solid plan for the essay. The best ideas from the brainstorming results can then be developed into a more formal outline. An outline typically has one main point (the thesis) and at least three sub-points that support the main point. Here's an example:

Main Idea

- Point #1
- Point #2
- Point #3

Of course, there will be details under each point, but this approach is the best for dealing with timed writing.

Staying on Track

Basing the essay on the outline aids in both organization and coherence. The goal is to ensure that there is enough time to develop each sub-point in the essay, roughly spending an equal amount of time on each idea. Keeping an eye on the time will help. If there are fifteen minutes left to draft the essay, then it makes sense to spend about 5 minutes on each of the ideas. Staying on task is critical to success, and timing out the parts of the essay can help writers avoid feeling overwhelmed.

Parts of the Essay

The introduction has to do a few important things:

- Establish the topic of the essay in original wording (i.e., not just repeating the prompt)

- Clarify the significance/importance of the topic or purpose for writing (not too many details, a brief overview)

- Offer a thesis statement that identifies the writer's own viewpoint on the topic (typically one-two brief sentences as a clear, concise explanation of the main point on the topic)

Body paragraphs reflect the ideas developed in the outline. Three-four points is probably sufficient for a short essay, and they should include the following:

- A topic sentence that identifies the sub-point (e.g., a reason why, a way how, a cause or effect)

- A detailed explanation of the point, explaining why the writer thinks this point is valid

- Illustrative examples, such as personal examples or real-world examples, that support and validate the point (i.e., "prove" the point)

- A concluding sentence that connects the examples, reasoning, and analysis to the point being made

The conclusion, or final paragraph, should be brief and should reiterate the focus, clarifying why the discussion is significant or important. It is important to avoid adding specific details or new ideas to this paragraph. The purpose of the conclusion is to sum up what has been said to bring the discussion to a close.

Don't Panic!

Writing an essay can be overwhelming, and performance panic is a natural response. The outline serves as a basis for the writing and helps writers keep focused. Getting stuck can also happen, and it's helpful to remember that brainstorming can be done at any time during the writing process. Following the steps of the writing process is the best defense against writer's block.

Timed essays can be particularly stressful, but assessors are trained to recognize the necessary planning and thinking for these timed efforts. Using the plan above and sticking to it helps with time management. Timing each part of the process helps writers stay on track. Sometimes writers try to cover too much in their essays. If time seems to be running out, this is an opportunity to determine whether all of the ideas in the outline are necessary. Three body paragraphs are sufficient, and more than that is probably too much to cover in a short essay.

More isn't always better in writing. A strong essay will be clear and concise. It will avoid unnecessary or repetitive details. It is better to have a concise, five-paragraph essay that makes a clear point, than a ten-paragraph essay that doesn't. The goal is to write one to two pages of quality writing. Paragraphs should also reflect balance; if the introduction goes to the bottom of the first page, the writing may be going off-track or be repetitive. It's best to fall into the one-two page range, but a complete, well-developed essay is the ultimate goal.

Math

1. If $4x - 3 = 5$, then $x =$
   a. 1
   b. 2
   c. 3
   d. 4

2. Write the expression for three times the sum of twice a number and one minus 6.
   a. $2x + 1 - 6$
   b. $3x + 1 - 6$
   c. $3(x + 1) - 6$
   d. $3(2x + 1) - 6$

3. On Monday, Robert mopped the floor in 4 hours. On Tuesday, he did it in 3 hours. If on Monday, his average rate of mopping was p sq. ft. per hour, what was his average rate on Tuesday?
   a. $\frac{4}{3}p$ sq. ft. per hour

   b. $\frac{3}{4}p$ sq. ft. per hour

   c. $\frac{5}{4}p$ sq. ft. per hour

   d. $p + 1$ sq. ft. per hour

4. Which of the following inequalities is equivalent to $3 - \frac{1}{2}x \geq 2$?
   a. $x \geq 2$
   b. $x \leq 2$
   c. $x \geq 1$
   d. $x \leq 1$

5. For which of the following are $x = 4$ and $x = -4$ solutions?
   a. $x^2 + 16 = 0$
   b. $x^2 + 4x - 4 = 0$
   c. $x^2 - 2x - 2 = 0$
   d. $x^2 - 16 = 0$

6. $(2x - 4y)^2 =$
   a. $4x^2 - 16xy + 16y^2$
   b. $4x^2 - 8xy + 16y^2$
   c. $4x^2 - 16xy - 16y^2$
   d. $2x^2 - 8xy + 8y^2$

7. The square and circle have the same center. The circle has a radius of *r*. What is the area of the shaded region?

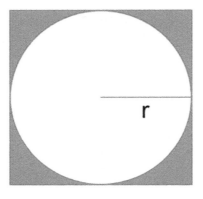

a. $r^2 - \pi r^2$
b. $4r^2 - 2\pi r$
c. $(4 - \pi)r^2$
d. $(\pi - 1)r^2$

8. Five of six numbers have a sum of 25. The average of all six numbers is 6. What is the sixth number?
   a. 8
   b. 10
   c. 11
   d. 12

9. What is the solution to the following system of equations?
$$x^2 - 2x + y = 8$$
$$x - y = -2$$
   a. $(-2, 3)$
   b. There is no solution.
   c. $(-2, 0) \, (1, 3)$
   d. $(-2, 0) \, (3, 5)$

10. Two cards are drawn from a shuffled deck of 52 cards. What's the probability that both cards are Kings if the first card isn't replaced after it's drawn?
   a. $\dfrac{1}{169}$

   b. $\dfrac{1}{221}$

   c. $\dfrac{1}{13}$

   d. $\dfrac{4}{13}$

11. The table below displays the number of three-year-olds at Kids First Daycare who are potty-trained and those who still wear diapers.

	Potty-trained	Wear diapers	
Boys	26	22	48
Girls	34	18	52
	60	40	

What is the probability that a three-year-old girl chosen at random from the school is potty-trained?
   a. 52 percent
   b. 34 percent
   c. 65 percent
   d. 57 percent

12. A shipping box has a length of 8 inches, a width of 14 inches, and a height of 4 inches. If all three dimensions are doubled, what is the relationship between the volume of the new box and the volume of the original box?
   a. The volume of the new box is double the volume of the original box.
   b. The volume of the new box is four times as large as the volume of the original box.
   c. The volume of the new box is six times as large as the volume of the original box.
   d. The volume of the new box is eight times as large as the volume of the original box.

13. Which of the following shows the correct result of simplifying the following expression:
$$(7n + 3n^3 + 3) + (8n + 5n^3 + 2n^4)$$
   a. $9n^4 + 15n - 2$
   b. $2n^4 + 5n^3 + 15n - 2$
   c. $9n^4 + 8n^3 + 15n$
   d. $2n^4 + 8n^3 + 15n + 3$

14. An equation for the line passing through the origin and the point $(2, 1)$ is
   a. $y = 2x$
   b. $y = \frac{1}{2}x$
   c. $y = x - 2$
   d. $2y = x + 1$

15. If $g(x) = x^3 - 3x^2 - 2x + 6$ and $f(x) = 2$, then what is $g(f(x))$?
   a. -26
   b. 6
   c. $2x^3 - 6x^2 - 4x + 12$
   d. -2

16. If the volume of a sphere is 288π cubic meters, what are the radius and surface area of the same sphere?
   a. Radius is 6 meters and surface area is 144π square meters
   b. Radius is 36 meters and surface area is 144π square meters
   c. Radius is 6 meters and surface area is 12π square meters
   d. Radius is 36 meters and surface area is 12π square meters

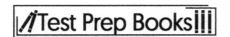

17. A ball is drawn at random from a ball pit containing 8 red balls, 7 yellow balls, 6 green balls, and 5 purple balls. What's the probability that the ball drawn is yellow?

    a. $^1/_{26}$

    b. $^{19}/_{26}$

    c. $^7/_{26}$

    d. 1

18. If Sarah reads at an average rate of 21 pages in four nights, how long will it take her to read 140 pages?

    a. 6 nights

    b. 26 nights

    c. 8 nights

    d. 27 nights

19. The phone bill is calculated each month using the equation $c = 50g + 75$. The cost of the phone bill per month is represented by c, and g represents the gigabytes of data used that month. Identify and interpret the slope of this equation.

    a. 75 dollars per day

    b. 75 gigabytes per day

    c. 50 dollars per day

    d. 50 dollars per gigabyte

20. For the following similar triangles, what are the values of x and y (rounded to one decimal place)?

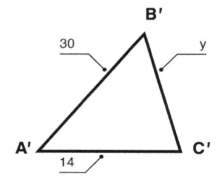

    a. $x = 16.5, y = 25.1$

    b. $x = 19.5, y = 24.1$

    c. $x = 17.1, y = 26.3$

    d. $x = 26.3, y = 17.1$

Reading

Passage #1

Questions 1-6 are based upon the following passage:

This excerpt is an adaptation of Jonathan Swift's *Gulliver's Travels into Several Remote Nations of the World.*

My gentleness and good behaviour had gained so far on the emperor and his court, and indeed upon the army and people in general, that I began to conceive hopes of getting my liberty in a short time. I took all possible methods to cultivate this favourable disposition. The natives came, by degrees, to be less apprehensive of any danger from me. I would sometimes lie down, and let five or six of them dance on my hand; and at last the boys and girls would venture to come and play at hide-and-seek in my hair. I had now made a good progress in understanding and speaking the language. The emperor had a mind one day to entertain me with several of the country shows, wherein they exceed all nations I have known, both for dexterity and magnificence. I was diverted with none so much as that of the rope-dancers, performed upon a slender white thread, extended about two feet, and twelve inches from the ground. Upon which I shall desire liberty, with the reader's patience, to enlarge a little.

This diversion is only practised by those persons who are candidates for great employments, and high favour at court. They are trained in this art from their youth, and are not always of noble birth, or liberal education. When a great office is vacant, either by death or disgrace (which often happens,) five or six of those candidates petition the emperor to entertain his majesty and the court with a dance on the rope; and whoever jumps the highest, without falling, succeeds in the office. Very often the chief ministers themselves are commanded to show their skill, and to convince the emperor that they have not lost their faculty. Flimnap, the treasurer, is allowed to cut a caper on the straight rope, at least an inch higher than any other lord in the whole empire. I have seen him do the summerset several times together, upon a trencher fixed on a rope which is no thicker than a common packthread in England. My friend Reldresal, principal secretary for private affairs, is, in my opinion, if I am not partial, the second after the treasurer; the rest of the great officers are much upon a par.

1. Which of the following statements best summarize the central purpose of this text?
   a. Gulliver details his fondness for the archaic yet interesting practices of his captors.
   b. Gulliver conjectures about the intentions of the aristocratic sector of society.
   c. Gulliver becomes acquainted with the people and practices of his new surroundings.
   d. Gulliver's differences cause him to become penitent around new acquaintances.

2. What is the word *principal* referring to in the following text?

> My friend Reldresal, principal secretary for private affairs, is, in my opinion, if I am not partial, the second after the treasurer; the rest of the great officers are much upon a par.

a. Primary or chief
b. An acolyte
c. An individual who provides nurturing
d. One in a subordinate position

3. What can the reader infer from this passage?

> I would sometimes lie down, and let five or six of them dance on my hand; and at last the boys and girls would venture to come and play at hide-and-seek in my hair.

a. The children tortured Gulliver.
b. Gulliver traveled because he wanted to meet new people.
c. Gulliver is considerably larger than the children who are playing around him.
d. Gulliver has a genuine love and enthusiasm for people of all sizes.

4. What is the significance of the word *mind* in the following passage?

> The emperor had a mind one day to entertain me with several of the country shows, wherein they exceed all nations I have known, both for dexterity and magnificence.

a. The ability to think
b. A collective vote
c. A definitive decision
d. A mythological question

5. Which of the following assertions does NOT support the fact that games are a commonplace event in this culture?

a. My gentlest and good behavior . . . short time.
b. They are trained in this art from their youth . . . liberal education.
c. Very often the chief ministers themselves are commanded to show their skill . . . not lost their faculty.
d. Flimnap, the treasurer, is allowed to cut a caper on the straight rope . . . higher than any other lord in the whole empire.

6. How do Flimnap and Reldresal demonstrate the community's emphasis on physical strength and leadership abilities?

a. Only children used Gulliver's hands as a playground.
b. The two men who exhibited superior abilities held prominent positions in the community.
c. Only common townspeople, not leaders, walk the straight rope.
d. No one could jump higher than Gulliver.

Passage #2

Questions 7-12 are based upon the following passage:

This excerpt is an adaptation of Robert Louis Stevenson's *The Strange Case of Dr. Jekyll and Mr. Hyde.*

"Did you ever come across a protégé of his—one Hyde?" He asked.

"Hyde?" repeated Lanyon. "No. Never heard of him. Since my time."

That was the amount of information that the lawyer carried back with him to the great, dark bed on which he tossed to and fro until the small hours of the morning began to grow large. It was a night of little ease to his toiling mind, toiling in mere darkness and besieged by questions.

Six o'clock struck on the bells of the church that was so conveniently near to Mr. Utterson's dwelling, and still he was digging at the problem. Hitherto it had touched him on the intellectual side alone; but now his imagination also was engaged, or rather enslaved; and as he lay and tossed in the gross darkness of the night in the curtained room, Mr. Enfield's tale went by before his mind in a scroll of lighted pictures. He would be aware of the great field of lamps in a nocturnal city; then of the figure of a man walking swiftly; then of a child running from the doctor's; and then these met, and that human Juggernaut trod the child down and passed on regardless of her screams. Or else he would see a room in a rich house, where his friend lay asleep, dreaming and smiling at his dreams; and then the door of that room would be opened, the curtains of the bed plucked apart, the sleeper recalled, and, lo! There would stand by his side a figure to whom power was given, and even at that dead hour he must rise and do its bidding. The figure in these two phrases haunted the lawyer all night; and if at anytime he dozed over, it was but to see it glide more stealthily through sleeping houses, or move the more swiftly, and still the more smoothly, even to dizziness, through wider labyrinths of lamplighted city, and at every street corner crush a child and leave her screaming. And still the figure had no face by which he might know it; even in his dreams it had no face, or one that baffled him and melted before his eyes; and thus there it was that there sprung up and grew apace in the lawyer's mind a singularly strong, almost an inordinate, curiosity to behold the features of the real Mr. Hyde. If he could but once set eyes on him, he thought the mystery would lighten and perhaps roll altogether away, as was the habit of mysterious things when well examined. He might see a reason for his friend's strange preference or bondage, and even for the startling clauses of the will. And at least it would be a face worth seeing: the face of a man who was without bowels of mercy: a face which had but to show itself to raise up, in the mind of the unimpressionable Enfield, a spirit of enduring hatred.

From that time forward, Mr. Utterson began to haunt the door in the by street of shops. In the morning before office hours, at noon when business was plenty of time scarce, at night under the face of the full city moon, by all lights and at all hours of solitude or concourse, the lawyer was to be found on his chosen post.

"If he be Mr. Hyde," he had thought, "I should be Mr. Seek."

7. What is the purpose of the use of repetition in the following passage?

> It was a night of little ease to his toiling mind, toiling in mere darkness and besieged by questions.

a. It serves as a demonstration of the mental state of Mr. Lanyon.
b. It is reminiscent of the church bells that are mentioned in the story.
c. It mimics Mr. Utterson's ambivalence.
d. It emphasizes Mr. Utterson's anguish in failing to identify Hyde's whereabouts.

8. What is the setting of the story in this passage?

a. In the city
b. On the countryside
c. In a jail
d. In a mental health facility

9. What can one infer about the meaning of the word "Juggernaut" from the author's use of it in the passage?

a. It is an apparition that appears at daybreak.
b. It scares children.
c. It is associated with space travel.
d. Mr. Utterson finds it soothing.

10. What is the definition of the word *haunt* in the following passage?

> From that time forward, Mr. Utterson began to haunt the door in the by street of shops. In the morning before office hours, at noon when business was plenty of time scarce, at night under the face of the full city moon, by all lights and at all hours of solitude or concourse, the lawyer was to be found on his chosen post.

a. To levitate
b. To constantly visit
c. To terrorize
d. To daunt

11. The phrase *labyrinths of lamplighted city* contains an example of what?

a. Hyperbole
b. Simile
c. Juxtaposition
d. Alliteration

12. What can one reasonably conclude from the final comment of this passage?

> "If he be Mr. Hyde," he had thought, "I should be Mr. Seek."

a. The speaker is considering a name change.
b. The speaker is experiencing an identity crisis.
c. The speaker has mistakenly been looking for the wrong person.
d. The speaker intends to continue to look for Hyde.

Passage #3

Questions 13-18 are based upon the following passage:

This excerpt is adaptation from Abraham Lincoln's Address Delivered at the Dedication of the Cemetery at Gettysburg, November 19, 1863.

> Four score and seven years ago our fathers brought forth on this continent, a new nation, conceived in liberty, and dedicated to the proposition that all men are created equal.
>
> Now we are engaged in a great civil war, testing whether that nation, or any nation so conceived and so dedicated, can long endure. We are met on a great battlefield of that war. We have come to dedicate a portion of that field, as a final resting place for those who here gave their lives that this nation might live. It is altogether fitting and proper that we should do this.
>
> But, in a larger sense, we cannot dedicate—we cannot consecrate that we cannot hallow—this ground. The brave men, living and dead, who struggled here, have consecrated it, far above our poor power to add or detract. The world will little note, nor long remember what we say here, but it can never forget what they did here. It is for us the living, rather, to be dedicated here to the unfinished work which they who fought here have thus far so nobly advanced. It is rather for us to be here and dedicated to the great task remaining before us—that from these honored dead we take increased devotion to that cause for which they gave the last full measure of devotion—that we here highly resolve that these dead shall not have died in vain—that these this nation, under God, shall have a new birth of freedom—and that government of people, by the people, for the people, shall not perish from the earth.

13. The best description for the phrase *four score and seven years ago* is which of the following?
    a. A unit of measurement
    b. A period of time
    c. A literary movement
    d. A statement of political reform

14. What is the setting of this text?
    a. A battleship off of the coast of France
    b. A desert plain on the Sahara Desert
    c. A battlefield in North America
    d. The residence of Abraham Lincoln

15. Which war is Abraham Lincoln referring to in the following passage?
    > Now we are engaged in a great civil war, testing whether that nation, or any nation so conceived and so dedicated, can long endure.

    a. World War I
    b. The War of the Spanish Succession
    c. World War II
    d. The American Civil War

16. What message is the author trying to convey through this address?
   a. The audience should perpetuate the ideals of freedom that the soldiers died fighting for.
   b. The audience should honor the dead by establishing an annual memorial service.
   c. The audience should form a militia that would overturn the current political structure.
   d. The audience should forget the lives that were lost and discredit the soldiers.

17. Which rhetorical device is being used in the following passage?
   . . . we here highly resolve that these dead shall not have died in vain—that these this nation, under God, shall have a new birth of freedom—and that government of people, by the people, for the people, shall not perish from the earth.

   a. Antimetabole
   b. Antiphrasis
   c. Anaphora
   d. Epiphora

18. What is the effect of Lincoln's statement in the following passage?
   But, in a larger sense, we cannot dedicate—we cannot consecrate that we cannot hallow—this ground. The brave men, living and dead, who struggled here, have consecrated it, far above our poor power to add or detract.

   a. His comparison emphasizes the great sacrifice of the soldiers who fought in the war.
   b. His comparison serves as a reminder of the inadequacies of his audience.
   c. His comparison serves as a catalyst for guilt and shame among audience members.
   d. His comparison attempts to illuminate the great differences between soldiers and civilians.

19. Annabelle Rice started having trouble sleeping. Her biological clock was suddenly amiss and she began to lead a nocturnal schedule. She thought her insomnia was due to spending nights writing a horror story, but then she realized that even the idea of going outside into the bright world scared her to bits. She concluded she was now suffering from heliophobia.

Which of the following most accurately describes the meaning of the underlined word in the sentence above?
   a. Fear of dreams
   b. Fear of sunlight
   c. Fear of strangers
   d. Anxiety spectrum disorder

20. Which of these descriptions would give the most detailed and objective support for the claim that drinking and driving is unsafe?
   a. A dramatized television commercial reenacting a fatal drinking and driving accident, including heart-wrenching testimonials from loved ones
   b. The Department of Transportation's press release noting the additional drinking and driving special patrol units that will be on the road during the holiday season
   c. Congressional written testimony on the number of drinking and driving incidents across the country and their relationship to underage drinking statistics, according to experts
   d. A highway bulletin warning drivers of the penalties associated with drinking and driving

21. In 2015, 28 countries, including Estonia, Portugal, Slovenia, and Latvia, scored significantly higher than the United States on standardized high school math tests. In the 1960s, the United States

consistently ranked first in the world. Today, the United States spends more than $800 billion dollars on education, which exceeds the next highest country by more than $600 billion dollars. The United States also leads the world in spending per school-aged child by an enormous margin.

If these statements above are factual, which of the following statements must be correct?
a. Outspending other countries on education has benefits beyond standardized math tests.
b. The United States' education system is corrupt and broken.
c. The standardized math tests are not representative of American academic prowess.
d. Spending more money does not guarantee success on standardized math tests.

22. Raul is going to Egypt next month. He has been looking forward to this vacation all year. Since childhood, Raul has been fascinated with pyramids, especially the Great Pyramid of Giza, which is the oldest of the Seven Wonders of the Ancient World. According to religious custom, Egyptian royalty is buried in the tombs located within the pyramid's great labyrinths. Since it has been many years since Raul read about the pyramid's history, he wants to read a book describing how and why the Egyptians built the Great Pyramid thousands of years ago.

Which of the following guides would be the best for Raul?
a. *A Beginner's Guide to Giza*, a short book describing the city's best historical sites, published by the Egyptian Tourism Bureau (2015)
b. *The Life of Zahi Hawass*, the autobiography of one of Egypt's most famous archaeologists who was one of the first explorers at Giza (2014)
c. *A History of Hieroglyphics*, an in-depth look at how archaeologists first broke the ancient code, published by the University of Giza's famed history department (2013)
d. *Who Built the Great Pyramids?*, a short summary of the latest research and theories on the ancient Egyptians' religious beliefs and archaeological skills, written by a team of leading experts in the field (2015)

Question 23 is based on the following passage:

Cynthia keeps to a strict vegetarian diet, which is part of her religion. She absolutely cannot have any meat or fish dishes. This is more than a preference; her body has never developed the enzymes to process meat or fish, so she becomes violently ill if she accidentally eats any of the offending foods.

Cynthia is attending a full day event at her college next week. When at an event that serves meals, she always likes to bring a platter of vegetarian food for herself and to share with other attendees who have similar dietary restrictions. She requested a menu in advance to determine when her platter might be most useful to vegetarians. Here is the menu:

Breakfast: Hazelnut coffee or English breakfast tea, French toast, eggs, and bacon strips

Lunch: Assorted sandwiches (vegetarian options available), French fries, and baked beans

Cocktail hour: Alcoholic beverages, fruit, and cheese

Dinner: Roasted pork loin, seared trout, and bacon-bit topped macaroni and cheese

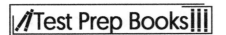

23. If Cynthia wants to pick the meal where there would be the least options for her and fellow vegetarians, during what meal should she bring the platter?
    a. Breakfast
    b. Lunch
    c. Cocktail hour
    d. Dinner

24. Technology has been invading cars for the last several years, but there are some new high tech trends that are pretty amazing. It is now standard in many car models to have a rear-view camera, hands-free phone and text, and a touch screen digital display. Music can be streamed from a paired cell phone, and some displays can even be programmed with a personal photo. Sensors beep to indicate there is something in the driver's path when reversing and changing lanes. Rain-sensing windshield wipers and lights are automatic, leaving the driver with little to do but watch the road and enjoy the ride. The next wave of technology will include cars that automatically parallel park, and a self-driving car is on the horizon. These technological advances make it a good time to be a driver.

It can be concluded from this paragraph that:
    a. Technology will continue to influence how cars are made.
    b. Windshield wipers and lights are always automatic.
    c. It is standard to have a rear-view camera in all cars.
    d. Technology has reached its peak in cars.

Writing

1. The following sentence contains what kind of error?
    > This summer, I'm planning to travel to Italy, take a Mediterranean cruise, going to Pompeii, and eat a lot of Italian food.

    a. Parallelism
    b. Sentence fragment
    c. Misplaced modifier
    d. Subject-verb agreement

2. The following sentence contains what kind of error?
    > Forgetting that he was supposed to meet his girlfriend for dinner, Anita was mad when Fred showed up late.

    a. Parallelism
    b. Run-on sentence
    c. Misplaced modifier
    d. Subject-verb agreement

3. The following sentence contains what kind of error?
    > Some workers use all their sick leave, other workers cash out their leave.

    a. Parallelism
    b. Comma splice
    c. Sentence fragment
    d. Subject-verb agreement

4. A student writes the following in an essay:
>   Protestors filled the streets of the city. Because they were dissatisfied with the government's
>   leadership.

Which of the following is an appropriately-punctuated correction for this sentence?
>   a. Protestors filled the streets of the city, because they were dissatisfied with the government's
>   leadership.
>   b. Protesters, filled the streets of the city, because they were dissatisfied with the government's
>   leadership.
>   c. Because they were dissatisfied with the government's leadership protestors filled the streets of
>   the city.
>   d. Protestors filled the streets of the city because they were dissatisfied with the government's
>   leadership.

5. Which word choices will correctly complete the sentence?
>   Increasing the price of bus fares has had a greater [affect / effect] on ridership [then / than]
>   expected.

>   a. affect; then
>   b. affect; than
>   c. effect; then
>   d. effect; than

Directions for questions 6–15

*Rewrite the sentence in your head following the directions given below. Keep in mind that your new
sentence should be well written and should have essentially the same meaning as the original sentence.*

6. Although she was nervous speaking in front of a crowd, the author read her narrative with poise and
confidence.

Rewrite, beginning with

The author had poise and confidence while reading

The next words will be
>   a. because she was nervous speaking in front of a crowd.
>   b. but she was nervous speaking in front of a crowd.
>   c. even though she was nervous speaking in front of a crowd.
>   d. before she was nervous speaking in front of a crowd.

7. There was a storm surge and loss of electricity during the hurricane.

Rewrite, beginning with

While the hurricane occurred,

The next words will be
   a. there was a storm surge after the electricity went out.
   b. the storm surge caused the electricity to go out.
   c. the electricity surged into the storm.
   d. the electricity went out, and there was a storm surge.

8. When one elephant in a herd is sick, the rest of the herd will help it walk and bring it food.

Rewrite, beginning with

<u>An elephant herd will</u>

The next words will be
   a. be too sick and tired to walk.
   b. help and support.
   c. gather food when they're sick.
   d. be unable to walk without food.

9. They went out to eat after the soccer game.

Rewrite, beginning with

<u>They finished the soccer game</u>

The next words will be
   a. then went out to eat.
   b. after they went out to eat.
   c. so they could go out to eat.
   d. because they went out to eat.

10. Armani got lost when she walked around Paris.

Rewrite, beginning with

<u>Walking through Paris,</u>

The next words will be
   a. you can get lost.
   b. Armani found herself lost.
   c. she should have gotten lost.
   d. is about getting lost.

11. After his cat died, Phoenix buried the cat with her favorite toys in his backyard.

Rewrite, beginning with

<u>Phoenix buried his cat</u>

The next words will be

a. in his backyard before she died.
b. after she died in the backyard.
c. with her favorite toys after she died.
d. after he buried her toys in the backyard.

12. While I was in the helicopter, I saw the sunset, and tears streamed down my eyes.

Rewrite, beginning with

<u>Tears streamed down my eyes</u>

The next words will be:
a. while I watched the helicopter fly into the sunset.
b. because the sunset flew up into the sky.
c. because the helicopter was facing the sunset.
d. when I saw the sunset from the helicopter.

13. I won't go to the party unless some of my friends go.

Rewrite, beginning with

<u>I will go the party</u>

The next words will be
a. if I want to.
b. if my friends go.
c. since a couple of my friends are going.
d. unless people I know go.

14. He had a broken leg before the car accident, so it took him a long time to recover.

Rewrite, beginning with

<u>He took a long time to recover from the car accident</u>

The next words will be
a. from his two broken legs.
b. after he broke his leg.
c. because he already had a broken leg.
d. since he broke his leg again afterward.

15. We had a party the day after Halloween to celebrate my birthday.

Rewrite, beginning with

<u>It was my birthday</u>

The next words will be
a. , so we celebrated with a party the day after Halloween.
b. the day of Halloween so we celebrated with a party.
c. , and we celebrated with a Halloween party the day after.
d. a few days before Halloween, so we threw a party.

Read the following passage and answer Questions 16-20.

1 Although many Missourians know that Harry S. Truman and Walt Disney hailed from their great state, probably far fewer know that it was also home to the remarkable George Washington Carver. (21) As a child, George was driven to learn, and he loved painting. At the end of the Civil War, Moses Carver, the slave owner who owned George's parents, decided to keep George and his brother and raise them on his farm.

2 He even went on to study art while in college but was encouraged to pursue botany instead. He spent much of his life helping others (22) by showing them better ways to farm, his ideas improved agricultural productivity in many countries. One of his most notable contributions to the newly emerging class of Black farmers was to teach them the negative effects of agricultural monoculture, i.e. (23) growing the same crops in the same fields year after year, depleting the soil of much needed nutrients and results in a lesser yielding crop.

3 Carver was an innovator, always thinking of new and better ways to do things, and is most famous for his over three hundred uses for the peanut. Toward the end of his career, (24) Carver returns to his first love of art. Through his artwork, he hoped to inspire people to see the beauty around them and to do great things themselves. (25) Because Carver died, he left his money to help fund ongoing agricultural research. Today, people still visit and study at the George Washington Carver Foundation at Tuskegee Institute.

16. Which of the following would be the best choice for this sentence (reproduced below)?

    As a child, George was driven to learn, and he loved painting.

    a. Leave it as it is now
    b. Move to the end of the first paragraph.
    c. Move to the beginning of the first paragraph.
    d. Move to the end of the second paragraph.

17. Which is the best version of the underlined portion of this sentence (reproduced below)?

    He spent much of his life helping others by showing them better ways to farm, his ideas improved agricultural productivity in many countries.

    a. (as it is now)
    b. by showing them better ways to farm his ideas improved agricultural productivity
    c. by showing them better ways to farm . . . his ideas improved agricultural productivity
    d. by showing them better ways to farm; his ideas improved agricultural productivity

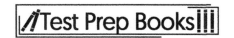
18. Which is the best version of the underlined portion of this sentence (reproduced below)?

One of his most notable contributions to the newly emerging class of Black farmers *was to teach them the negative effects of agricultural monoculture, i.e. growing the same crops in the same fields year after year, depleting the soil of much needed nutrients and results in a lesser yielding crop.*

a. (as it is now)
b. growing the same crops in the same fields year after year, depleting the soil of much needed nutrients and resulting in a lesser yielding crop.
c. growing the same crops in the same fields year after year, depletes the soil of much needed nutrients and resulting in a lesser yielding crop.
d. grows the same crops in the same fields year after year, depletes the soil of much needed nutrients and resulting in a lesser yielding crop.

19. In context, which is the best version of the underlined portion of this sentence (reproduced below)?

Toward the end of his career, *Carver returns* to his first love of art.

a. (as it is now)
b. Carver is returning
c. Carver returned
d. Carver was returning

20. In context, which is the best version of the underlined portion of this sentence (reproduced below)?

Because Carver died, he left his money to help fund ongoing agricultural research.

a. (as it is now)
b. Although Carver died,
c. When Carver died,
d. Finally Carver died,

Essay

Prepare an essay of about 300-600 words on the topic below.

Some people feel that sharing their lives on social media sites such as Facebook, Instagram, and Snapchat is fine. They share every aspect of their lives, including pictures of themselves and their families, what they ate for lunch, who they are dating, and when they are going on vacation. They even say that if it's not on social media, it didn't happen. Other people believe that sharing so much personal information is an invasion of privacy and could prove dangerous. They think sharing personal pictures and details invites predators, cyberbullying, and identity theft.

Write an essay to someone who is considering whether to participate in social media. Take a side on the issue and argue whether or not he/she should join a social media network. Use specific examples to support your argument.

Answer Explanations for Practice Test #1

Math

1. B: Add 3 to both sides to get $4x = 8$. Then divide both sides by 4 to get $x = 2$.

2. D: The expression is three times the sum of twice a number and 1, which is $3(2x + 1)$. Then, 6 is subtracted from this expression.

3. A: Robert accomplished his task on Tuesday in $\frac{3}{4}$ the time compared to Monday. He must have worked $\frac{4}{3}$ as fast.

4. B: To simplify this inequality, subtract 3 from both sides to get $-\frac{1}{2}x \geq -1$. Then, multiply both sides by -2 (remembering this flips the direction of the inequality) to get $x \leq 2$.

5. D: There are two ways to approach this problem. Each value can be substituted into each equation. Choice *A* can be eliminated, since $4^2 + 16 = 32$. Choice *B* can be eliminated, since $4^2 + 4 \times 4 - 4 = 28$. Choice *C* can be eliminated, since $4^2 - 2 \times 4 - 2 = 6$. But, plugging in either value into $x^2 - 16$, which gives $(\pm 4)^2 - 16 = 16 - 16 = 0$.

6. A: To expand a squared binomial, it's necessary to use the *First, Inner, Outer, Last Method*.

$$(2x - 4y)^2$$

$$2x \cdot 2x + 2x(-4y) + (-4y)(2x) + (-4y)(-4y)$$

$$4x^2 - 8xy - 8xy + 16y^2$$

$$y^2 4x^2 - 16xy + 16y^2$$

7. C: The area of the shaded region is the area of the square, minus the area of the circle. The area of the circle will be πr^2. The side of the square will be $2r$, so the area of the square will be $4r^2$. Therefore, the difference is $4r^2 - \pi r^2 = (4 - \pi)r^2$.

8. C: The average is calculated by adding all six numbers, then dividing by 6. The first five numbers have a sum of 25. If the total divided by 6 is equal to 6, then the total itself must be 36. The sixth number must be $36 - 25 = 11$.

9. D: This system of equations involves one quadratic function and one linear function, as seen from the degree of each equation. One way to solve this is through substitution. Solving for y in the second equation yields $y = x + 2$. Plugging this equation in for the y of the quadratic equation yields $x^2 - 2x + x + 2 = 8$. Simplifying the equation, it becomes $x^2 - x + 2 = 8$. Setting this equal to zero and factoring, it becomes $x^2 - x - 6 = 0 = (x - 3)(x + 2)$. Solving these two factors for x gives the zeros $x = 3, -2$. To find the y-value for the point, each number can be plugged in to either original equation. Solving each one for y yields the points $(3, 5)$ and $(-2, 0)$.

10. B: For the first card drawn, the probability of a King being pulled is $\frac{4}{52}$. Since this card isn't replaced, if a King is drawn first the probability of a King being drawn second is $\frac{3}{51}$. The probability of a King being drawn in both the first and second draw is the product of the two probabilities: $\frac{4}{52} \times \frac{3}{51} = \frac{12}{2652}$. This fraction, when divided by 12, equals $\frac{1}{221}$.

11. C: The conditional frequency of a girl being potty-trained is calculated by dividing the number of potty-trained girls by the total number of girls: $34 \div 52 = 0.65$. To determine the conditional probability, multiply the conditional frequency by 100: $0.65 \times 100 = 65\%$.

12. D: The formula for finding the volume of a rectangular prism is $V = l \times w \times h$ where l is the length, w is the width, and h is the height. The volume of the original box is calculated: $V = 8 \times 14 \times 4 = 448 \text{ in}^3$. The volume of the new box is calculated: $V = 16 \times 28 \times 8 = 3584 \text{ in}^3$. The volume of the new box divided by the volume of the old box equals 8.

13. D: The expression is simplified by collecting like terms. Terms with the same variable and exponent are like terms, and their coefficients can be added.

14. B: The slope will be given by $\frac{1-0}{2-0} = \frac{1}{2}$. The y-intercept will be 0, since it passes through the origin. Using slope-intercept form, the equation for this line is $y = \frac{1}{2}x$.

15. D: This problem involves a composition function, where one function is plugged into the other function. In this case, the $f(x)$ function is plugged into the $g(x)$ function for each x-value. The composition equation becomes $g\big(f(x)\big) = 2^3 - 3(2^2) - 2(2) + 6$. Simplifying the equation gives the answer $g\big(f(x)\big) = 8 - 3(4) - 2(2) + 6 = 8 - 12 - 4 + 6 = -2$.

16. A: Because the volume of the given sphere is 288π cubic meters, this means $\frac{4}{3}\pi r^3 = 288\pi$. This equation is solved for r to obtain a radius of 6 meters. The formula for the surface area of a sphere is $4\pi r^2$, so if $r = 6$ in this formula, the surface area is 144π square meters.

17. C: The sample space is made up of $8 + 7 + 6 + 5 = 26$ balls. The probability of pulling each individual ball is $\frac{1}{26}$. Since there are 7 yellow balls, the probability of pulling a yellow ball is $\frac{7}{26}$.

18. D: This problem can be solved by setting up a proportion involving the given information and the unknown value. The proportion is $\frac{21\ pages}{4\ nights} = \frac{140\ pages}{x\ nights}$. Solving the proportion by cross-multiplying, the equation becomes $21x = 4 \times 140$, where $x = 26.67$. Since it is not an exact number of nights, the answer is rounded up to 27 nights. Twenty-six nights would not give Sarah enough time.

19. D: The slope from this equation is 50, and it is interpreted as the cost per gigabyte used. Since the g-value represents number of gigabytes and the equation is set equal to the cost in dollars, the slope relates these two values. For every gigabyte used on the phone, the bill goes up 50 dollars.

20. C: Because the triangles are similar, the lengths of the corresponding sides are proportional. Therefore:

$$\frac{30 + x}{30} = \frac{22}{14} = \frac{y + 15}{y}$$

This results in the equation $14(30 + x) = 22 \cdot 30$ which, when solved, gives $x = 17.1$. The proportion also results in the equation $14(y + 15) = 22y$ which, when solved, gives $y = 26.3$.

Reading

1. C: Gulliver becomes acquainted with the people and practices of his new surroundings. Choice *C* is the correct answer because it most extensively summarizes the entire passage. While Choices *A* and *B* are reasonable possibilities, they reference portions of Gulliver's experiences, not the whole. Choice *D* is incorrect because Gulliver doesn't express repentance or sorrow in this particular passage.

2. A: Principal refers to *chief* or *primary* within the context of this text. Choice *A* is the answer that most closely aligns with this definition. Choices *B* and *D* make reference to a helper or followers while Choice *C* doesn't meet the description of Reldresal from the passage.

3. C: One can reasonably infer that Gulliver is considerably larger than the children who were playing around him because multiple children could fit into his hand. Choice *B* is incorrect because there is no indication of stress in Gulliver's tone. Choices *A* and *D* aren't the best answer because though Gulliver seems fond of his new acquaintances, he didn't travel there with the intentions of meeting new people or to express a definite love for them in this particular portion of the text.

4. C: The emperor made a *definitive decision* to expose Gulliver to their native customs. In this instance, the word *mind* was not related to a vote, question, or cognitive ability.

5. A: Choice *A* is correct. This assertion does not support the fact that games are a commonplace event in this culture because it mentions conduct, not games. Choices *B*, *C*, and *D* are incorrect because these do support the fact that games were a commonplace event.

6. B: Choice *B* is the only option that mentions the correlation between physical ability and leadership positions. Choices *A* and *D* are unrelated to physical strength and leadership abilities. Choice *C* does not make a deduction that would lead to the correct answer—it only comments upon the abilities of common townspeople.

7. D: It emphasizes Mr. Utterson's anguish in failing to identify Hyde's whereabouts. Context clues indicate that Choice *D* is correct because the passage provides great detail of Mr. Utterson's feelings about locating Hyde. Choice *A* does not fit because there is no mention of Mr. Lanyon's mental state. Choice *B* is incorrect; although the text does make mention of bells, Choice *B* is not the *best* answer overall. Choice *C* is incorrect because the passage clearly states that Mr. Utterson was determined, not unsure.

8. A: In the city. The word *city* appears in the passage several times, thus establishing the location for the reader.

9. B: It scares children. The passage states that the Juggernaut causes the children to scream. Choices *A* and *D* don't apply because the text doesn't mention either of these instances specifically. Choice *C* is incorrect because there is nothing in the text that mentions space travel.

10. B: To constantly visit. The mention of *morning*, *noon*, and *night* make it clear that the word *haunt* refers to frequent appearances at various locations. Choice *A* doesn't work because the text makes no mention of levitating. Choices *C* and *D* are not correct because the text makes mention of Mr. Utterson's anguish and disheartenment because of his failure to find Hyde but does not make mention of Mr. Utterson's feelings negatively affecting anyone else.

11. D: This is an example of alliteration. Choice *D* is the correct answer because of the repetition of the *L*-words. Hyperbole is an exaggeration, so Choice *A* doesn't work. No comparison is being made, so no simile or juxtaposition is being used, thus eliminating Choices *B* and *C*.

12. D: The speaker intends to continue to look for Hyde. Choices *A* and *B* are not possible answers because the text doesn't refer to any name changes or an identity crisis, despite Mr. Utterson's extreme obsession with finding Hyde. The text also makes no mention of a mistaken identity when referring to Hyde, so Choice *C* is also incorrect.

13. B: A period of time. It is apparent that Lincoln is referring to a period of time within the context of the passage because of how the sentence is structured with the word *ago*.

14. C: Lincoln's reference to *the brave men, living and dead, who struggled here,* proves that he is referring to a battlefield. Choices *A* and *B* are incorrect, as a *civil war* is mentioned and not a war with France or a war in the Sahara Desert. Choice *D* is incorrect because it does not make sense to consecrate a President's ground instead of a battlefield ground for soldiers who died during the American Civil War.

15. D: Abraham Lincoln is the former president of the United States, and he references a "civil war" during his address.

16. A: The audience should perpetuate the ideals of freedom that the soldiers died fighting for. Lincoln doesn't address any of the topics outlined in Choices *B*, *C*, or *D*. Therefore, Choice *A* is the correct answer.

17. D: Choice *D* is the correct answer because of the repetition of the word *people* at the end of the passage. Choice *A*, antimetabole, is the repetition of words in a phrase or clause but in reverse order, such as: "I do what I like, and like what I do." Choice *B*, *antiphrasis*, is a form of denial of an assertion in a text. Choice *C*, *anaphora*, is the repetition that occurs at the beginning of sentences.

18. A: Choice *A* is correct because Lincoln's intention was to memorialize the soldiers who had fallen as a result of war as well as celebrate those who had put their lives in danger for the sake of their country. Choices *B* and *D* are incorrect because Lincoln's speech was supposed to foster a sense of pride among the members of the audience while connecting them to the soldiers' experiences.

19. B: The passage indicates that Annabelle has a fear of going outside into the daylight. Thus, *heliophobia* must refer to a fear of bright lights or sunlight. Choice *B* is the only answer that describes this.

20. C: The answer we seek has both the most detailed and objective information. Choice *A* describing a television commercial with a dramatized reenactment is not particularly detailed. Choice *B*, a notice to the public informing them of additional drinking and driving units on patrol, is not detailed and objective information. Choice *D*, a highway bulletin, does not present the type of information required. Choice *C* is the correct answer. The number of incidents and their relationship to a possible cause are both detailed and objective information.

21. D: Outspending other countries on education could have other benefits, but there is no reference to this in the passage, so Choice *A* is incorrect. Choice *B* is incorrect because the author does not mention corruption. Choice *C* is incorrect because there is nothing in the passage stating that the tests are not genuinely representative. Choice *D* is accurate because spending more money has not brought success.

The United States already spends the most money, and the country is not excelling on these tests. Choice *D* is the correct answer.

22. D: Raul wants a book that describes how and why ancient Egyptians built the Great Pyramid of Giza. Choice *A* is incorrect because it focuses more generally on Giza as a whole, rather than the Great Pyramid itself. Choice *B* is close but incorrect because it is an autobiography that will largely focus on the archaeologist's life. Choice *C* is wrong because it focuses on hieroglyphics; it is not directly on point. Choice *D*, the book directly covering the building of the Great Pyramids, should be most helpful.

23. D: Cynthia needs to select the meal with the least vegetarian options. Although the breakfast menu, Choice *A*, includes bacon, there is also coffee, tea, French toast, and eggs available. Choice *B*, lunch, includes an option for vegetarian sandwiches along with the French fries and baked beans. The cocktail hour, Choice *C*, does not contain meat or fish. In contrast, the dinner is a vegetarian's nightmare: nothing suitable is offered. Thus, dinner, Choice *D*, is the best answer.

24. A: The passage discusses recent technological advances in cars and suggests that this trend will continue in the future with self-driving cars. Choice *B* and *C* are not true, so these are both incorrect. Choice *D* is also incorrect because the passage suggests continuing growth in technology, not a peak.

Writing

1. A: Parallelism refers to consistent use of sentence structure or word form. In this case, the list within the sentence does not utilize parallelism; three of the verbs appear in their base form—*travel*, *take*, and *eat*—but one appears as a gerund—*going*. A parallel version of this sentence would be "This summer, I'm planning to travel to Italy, take a Mediterranean cruise, go to Pompeii, and eat a lot of Italian food." Choice *B* is incorrect because this description is a complete sentence. Choice *C* is incorrect as a misplaced modifier is a modifier that is not located appropriately in relation to the word or words they modify. Choice *D* is incorrect because subject-verb agreement refers to the appropriate conjugation of a verb in relation to its subject.

2. C: In this sentence, the modifier is the phrase "Forgetting that he was supposed to meet his girlfriend for dinner." This phrase offers information about Fred's actions, but the noun that immediately follows it is Anita, creating some confusion about the "do-er" of the phrase. A more appropriate sentence arrangement would be "Forgetting that he was supposed to meet his girlfriend for dinner, Fred made Anita mad when he showed up late." Choice *A* is incorrect as parallelism refers to the consistent use of sentence structure and verb tense, and this sentence is appropriately consistent. Choice *B* is incorrect as a run-on sentence does not contain appropriate punctuation for the number of independent clauses presented, which is not true of this description. Choice *D* is incorrect because subject-verb agreement refers to the appropriate conjugation of a verb relative to the subject, and all verbs have been properly conjugated.

3. B: A comma splice occurs when a comma is used to join two independent clauses together without the additional use of an appropriate conjunction. One way to remedy this problem is to replace the comma with a semicolon. Another solution is to add a conjunction: "Some workers use all their sick leave, but other workers cash out their leave." Choice *A* is incorrect as parallelism refers to the consistent use of sentence structure and verb tense; all tenses and structures in this sentence are consistent. Choice *C* is incorrect because a sentence fragment is a phrase or clause that cannot stand alone—this sentence contains two independent clauses. Choice *D* is incorrect because subject-verb

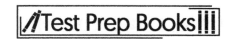

agreement refers to the proper conjugation of a verb relative to the subject, and all verbs have been properly conjugated.

4. D: The problem in the original passage is that the second sentence is a dependent clause that cannot stand alone as a sentence; it must be attached to the main clause found in the first sentence. Because the main clause comes first, it does not need to be separated by a comma. However, if the dependent clause came first, then a comma would be necessary, which is why Choice *C* is incorrect. *A* and *B* also insert unnecessary commas into the sentence.

5. D: In this sentence, the first answer choice requires a noun meaning *impact* or *influence*, so *effect* is the correct answer. For the second answer choice, the sentence is drawing a comparison. *Than* shows a comparative relationship whereas *then* shows sequence or consequence. Choices *A* and *C* can be eliminated because they contain the choice *then*. Choice *B* is incorrect because *affect* is a verb while this sentence requires a noun.

6. C: The original sentence states that despite the author being nervous, she was able to read with poise and confidence, which is stated in Choice *C*. Choice *A* changes the meaning by adding *because*; however, the author didn't read with confidence *because* she was nervous, but *despite* being nervous. Choice *B* is closer to the original meaning; however, it loses the emphasis of her succeeding *despite* her condition. Choice *D* adds the word *before*, which doesn't make much sense on its own, much less in relation to the original sentence.

7. D: The original sentence states that there was a storm surge and loss of electricity during the hurricane, making Choice *D* correct. Choices *A* and *B* arrange the storm surge and the loss of electricity within a cause and effect statement, which changes the meaning of the original sentence. Choice *C* changes *surge* from a noun into a verb and creates an entirely different situation.

8. B: The original sentence states that an elephant herd will help and support another herd member if it is sick, so Choice *B* is correct. Choice *A* is incorrect because it states the whole herd will be too sick and too tired to walk instead of a single elephant. Choice *C* is incorrect because the original sentence does not say that the herd gathers food when *they* are sick, but when a single member of the herd is sick. Although Choice *D* might be correct in a general sense, it does not relate to the meaning of the original sentence and is therefore incorrect.

9. A: The original sentence says that after a soccer game, they went out to eat. Choice *A* shows the same sequence: they finished the soccer game *then* went out to eat. Choice *B* is incorrect because it reverses the sequence of events. Choices *C* and *D* are incorrect because the words *so* and *because* change the meaning of the original sentence.

10. B: Choice *B* is correct because the idea of the original sentences is Armani getting lost while walking through Paris. Choice *A* is incorrect because it replaces third person with second person. Choice *C* is incorrect because the word *should* indicates an obligation to get lost. Choice *D* is incorrect because it is not specific to the original sentence but instead makes a generalization about getting lost.

11. C: Choice *C* is correct because it shows that Phoenix buried his cat with her favorite toys after she died, which is true of the original statement. Although Choices *A*, *B*, and *D* mention a backyard, the meanings of these choices are skewed. Choice *A* says that Phoenix buried his cat alive, which is incorrect. Choice *B* says his cat died in the backyard, which we do not know to be true. Choice *D* says Phoenix buried his cat after he buried her toys, which is also incorrect.

12. D: Choice *D* is correct because it expresses the sentiment of a moment of joy bringing tears to one's eyes as one sees a sunset while in a helicopter. Choice *A* is incorrect because it implies that the person was outside of the helicopter watching it from afar. Choice *B* is incorrect because the original sentence does not portray the sunset *flying up* into the sky. Choice *C* is incorrect because, while the helicopter may have been facing the sunset, this is not the reason that tears were in the speaker's eyes.

13. B: Choice *B* is correct because like the original sentence, it expresses their plan to go to the party if friends also go. Choice *A* is incorrect because it does not follow the meaning of the original sentence. Choice *C* is incorrect because it states that their friends are going, even though that is not known. Choice *D* is incorrect because it would make the new sentence mean the opposite of the original sentence.

14. C: Choice *C* is correct because the original sentence states that his recovery time was long because his leg was broken before the accident. Choice *A* is incorrect because there is no indication that the man had two broken legs. Choice *B* is incorrect because it indicates that he broke his leg during the car accident, not before. Choice *D* is incorrect because there is no indication that he broke his leg after the car accident.

15. A: Choice *A* is correct because it expresses the fact that the birthday and the party were both after Halloween. Choice *B* is incorrect because it says that the birthday was on Halloween, even though that was not stated in the original sentence. Choice *C* is incorrect because it says the party was specifically a Halloween party and not a birthday party. Choice *D* is incorrect because the party was after Halloween, not before.

16. B: The best place for this sentence given all the answer choices is at the end of the first paragraph. Choice *A* is incorrect; the passage is told in chronological order and leaving the sentence as-is defies that order, since we haven't been introduced to who raised George. Choice *C* is incorrect because this sentence is not an introductory sentence. It does not provide the main topic of the paragraph. Choice *D* is incorrect because again, it defies chronological order. By the end of paragraph two we have already gotten to George as an adult, so this sentence would not make sense here.

17. D: Out of these choices, a semicolon would be the best fit because there is an independent clause on either side of the semicolon, and the two sentences closely relate to each other. Choice *A* is incorrect because putting a comma between two independent clauses (i.e. complete sentences) creates a comma splice. Choice *B* is incorrect; omitting punctuation here creates a run-on sentence. Choice *C* is incorrect because an ellipses (. . .) is used to designate an omission in the text.

18. B: Choice *B* is the correct answer because it uses "ing" verbs as gerunds. Gerunds are "ing" words that stand in for nouns. The words "growing" and "depleting" are gerunds in this example. Choice *B* also uses the conjunction "and," whereas the other answer choices have comma splices.

19. C: Choice *C* is correct because it keeps with the verb tense in the rest of the passage: past tense. Choice *A* is in present tense, which is incorrect. Choice *B* is present progressive, which means there is a continual action, which is also incorrect. Choice *D* is incorrect because "was returning" is past progressive tense, which means that something was happening continuously at some point in the past.

20. C: The correct choice is the subordinating conjunction, "When." We should look at the clues around the phrase to see what fits best. Carver left his money "when he died." Choice *A*, "Because," could perhaps be correct, but "When" is the more appropriate word to use here. Choice *B* is incorrect; "Although" denotes a contrast, and there is no contrast here. Choice *D* is incorrect because "Finally" indicates something at the very end of a list or series, and there is no series at this point in the text.

TSI Practice Test #2

Math

1. Solve for x, if $x^2 - 2x - 8 = 0$.
   a. $2 \pm \frac{\sqrt{30}}{2}$
   b. $2 \pm 4\sqrt{2}$
   c. 1 ± 3
   d. $4 \pm \sqrt{2}$

2. Which graph will be a line parallel to the graph of $y = 3x - 2$?
   a. $2y - 6x = 2$
   b. $y - 4x = 4$
   c. $3y = x - 2$
   d. $2x - 2y = 2$

3. Jessica buys 10 cans of paint. Red paint costs $1 per can and blue paint costs $2 per can. In total, she spends $16. How many red cans did she buy?
   a. 2
   b. 3
   c. 4
   d. 5

4. For a group of 20 men, the median weight is 180 pounds and the range is 30 pounds. If each man gains 10 pounds, which of the following would be true?
   a. The median weight will increase, and the range will remain the same.
   b. The median weight and range will both remain the same.
   c. The median weight will stay the same, and the range will increase.
   d. The median weight and range will both increase.

5. A root of $x^2 - 2x - 2$ is
   a. $1 + \sqrt{3}$
   b. $1 + 2\sqrt{2}$
   c. $2 + 2\sqrt{3}$
   d. $2 - 2\sqrt{3}$

6. What is the product of the following expression?

$$(4x - 8)(5x^2 + x + 6)$$

   a. $20x^3 - 36x^2 + 16x - 48$
   b. $6x^3 - 41x^2 + 12x + 15$
   c. $204 + 11x^2 - 37x - 12$
   d. $2x^3 - 11x^2 - 32x + 20$

7. The graph shows the position of a car over a 10-second time interval. Which of the following is the correct interpretation of the graph for the interval 1 to 3 seconds?

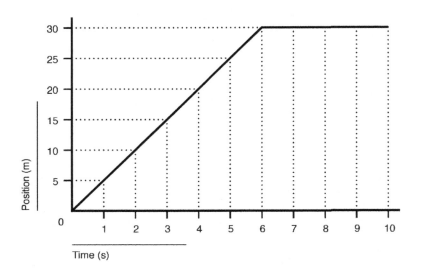

Time (s)

    a. The car remains in the same position.
    b. The car is traveling at a speed of 5 m/s.
    c. The car is traveling up a hill.
    d. The car is traveling at 5 mph.

8. What is the y-intercept for $y = x^2 + 3x - 4$?
    a. $y = 1$
    b. $y = -4$
    c. $y = 3$
    d. $y = 4$

9. What is the value of b in the equation: $5b - 4 = 2b + 17$?
    a. 13
    b. 24
    c. 7
    d. 21

10. Dwayne has received the following scores on his math tests: 78, 92, 83, and 97. What score must Dwayne get on his next math test to have an overall average of 90?
    a. 89
    b. 98
    c. 95
    d. 100

11. The following graph compares the various test scores of the top three students in each of these teacher's classes. Based on the graph, which teacher's students' test scores had the smallest range?

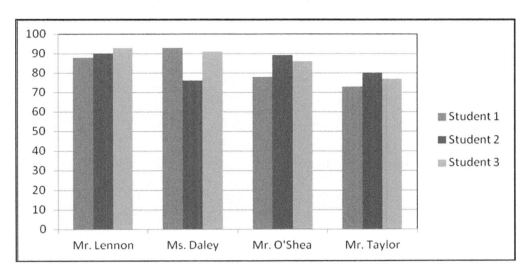

a. Mr. Lennon
b. Mr. O'Shea
c. Mr. Taylor
d. Ms. Daley

12. What's the probability of rolling a 6 at least once in two rolls of a die?
a. $1/3$
b. $1/36$
c. $1/6$
d. $11/36$

13. If the ordered pair $(-3, -4)$ is reflected over the x-axis, what's the new ordered pair?
a. $(-3, -4)$
b. $(3, -4)$
c. $(3, 4)$
d. $(-3, 4)$

14. If $-3(x + 4) \geq x + 8$, what is the value of x?
a. $x = 4$
b. $x \geq 2$
c. $x \geq -5$
d. $x \leq -5$

15. Karen gets paid a weekly salary and a commission for every sale that she makes. The table below shows the number of sales and her pay for different weeks.

Sales	2	7	4	8
Pay	$380	$580	$460	$620

Which of the following equations represents Karen's weekly pay?
   a. $y = 90x + 200$
   b. $y = 90x - 200$
   c. $y = 40x + 300$
   d. $y = 40x - 300$

16. Which of the ordered pairs below is a solution to the following system of inequalities?
$$y > 2x - 3$$
$$y < -4x + 8$$

   a. (4, 5)
   b. (−3, −2)
   c. (3, −1)
   d. (5, 2)

17. The area of a given rectangle is 24 square centimeters. If the measure of each side is multiplied by 3, what is the area of the new figure?
   a. 48 cm^2
   b. 72 cm^2
   c. 216 cm^2
   d. 13,824 cm^2

18. What is the perimeter of the figure below? Note that the solid outer line is the perimeter.

   a. 48.565 in
   b. 36.565 in
   c. 39.78 in
   d. 39.565 in

19. The graph of which function has an x-intercept of −2?
   a. $y = 2x - 3$
   b. $y = 4x + 2$
   c. $y = x^2 + 5x + 6$
   d. $y = -\frac{1}{2} \times 2^x$

20. In Jim's school, there are 3 girls for every 2 boys. There are 650 students in total. Using this information, how many students are girls?

    a. 260
    b. 130
    c. 65
    d. 390

Reading

Questions 1-4 are based on the following two passages:

Passage A

Excerpt from *Preface to Lyrical Ballads* by William Wordsworth (1800)

From such verses the Poems in these volumes will be found distinguished at least by one mark of difference, that each of them has a worthy *purpose*. Not that I always began to write with a distinct purpose formerly conceived; but habits of meditation have, I trust, so prompted and regulated my feelings, that my descriptions of such objects as strongly excite those feelings, will be found to carry along with them a *purpose.* If this opinion be erroneous, I can have little right to the name of a Poet. For all good poetry is the spontaneous overflow of powerful feelings: and though this be true, Poems to which any value can be attached were never produced on any variety of subjects but by a man who, being possessed of more than usual organic sensibility, had also thought long and deeply. For our continued influxes of feeling are modified and directed by our thoughts, which are indeed the representatives of all our past feelings; and, as by contemplating the relation of these general representatives to each other, we discover what is really important to men, so, by the repetition and continuance of this act, our feelings will be connected with important subjects, till at length, if we be originally possessed of much sensibility, such habits of mind will be produced, that, by obeying blindly and mechanically the impulses of those habits, we shall describe objects, and utter sentiments, of such a nature, and in such connexion with each other, that the understanding of the Reader must necessarily be in some degree enlightened, and his affections strengthened and purified.

Passage B

Excerpt from Tradition and the Individual Talent by T.S. Eliot (1921)

If you compare several representative passages of the greatest poetry you see how great is the variety of types of combination, and also how completely any semi-ethical criterion of "sublimity" misses the mark. For it is not the "greatness," the intensity, of the emotions, the components, but the intensity of the artistic process, the pressure, so to speak, under which the fusion takes place, that counts. The episode of Paolo and Francesca employs a definite emotion, but the intensity of the poetry is something quite different from whatever intensity in the supposed experience it may give the impression of. It is no more intense, furthermore, than Canto XXVI, the voyage of Ulysses, which has not the direct dependence upon an emotion. Great variety is possible in the process of transmution of emotion: the murder of Agamemnon, or the agony of Othello, gives an artistic effect apparently closer to a possible original than the scenes from Dante. In the *Agamemnon,* the artistic emotion approximates to the emotion of an actual spectator; in *Othello* to the emotion of the protagonist himself. But the difference between art and the event is always absolute; the combination which is the murder of Agamemnon is probably as complex as that which is the voyage of Ulysses. In either case there has been a fusion of elements. The ode of Keats contains a number of feelings which have nothing particular to do with the nightingale, but which the nightingale, partly, perhaps, because of its attractive name, and partly because of its reputation, served to bring together.

1. Which one of the following most accurately characterizes the relationship between the two passages?
   a. Passage A offers an explanation of the purpose of poetry, while passage B offers an explanation of the results of poetry.
   b. Passage A is concerned with the context of poetry involving nature, while passage B is concerned with the context of poetry involving urban life.
   c. Passage A focuses on lyric poetry, while passage B is concerned with epic poetry.
   d. Passage A argues that the source of great poetry comes from emotions within, while passage B argues that great poetry is based off the skill and expertise of the poet.

2. What does the author of passage B mean by the last sentence?
   a. The author is explaining that the feelings portrayed in Keats' ode are brought forth and signified by the symbolism of the nightingale used within the language of the poem.
   b. The author means that Keats' poetry is more valuable than epic verse because the feelings expressed in his poetry are backed by literary talent.
   c. The author means that the figure of the nightingale serves a myriad of purposes; here, for a symbolic use of a range of emotions such as grief, fear, and sorrow.
   d. The author is trying to argue that the reputation of the poet serves to bring about the feelings of the reader; or, that the ethos of the writing is closely related to the pathos in the audience.

3. The author of passage A thinks which of the following about thoughts and feelings as they relate to poetry?
   a. That the thoughts of the poet are the primary source of inspiration, and that feeling is a secondary motivation for the poet. Once the poet gets their thoughts onto paper, the feelings of the poet will come after.
   b. That thoughts and feelings are one and the same thing; that they happen simultaneously, and that this merging creates the perfect condition to write a poem.
   c. That feelings from the past will accumulate into representations of thought. The poet is then responsible for putting these direct thoughts onto paper, creating a poem. Therefore, the original source of poetry is from the past feelings of the poet.
   d. That thoughts and feelings are overrated; namely, that critics of the past have put too much emphasis on thoughts and feelings as they relate to poetry.

4. The authors of the passages differ in their explanations of great poetry in that the author of passage B
   a. argues that the intensity of the emotion felt while writing the poem counts for more than the effort put into a poem.
   b. argues that the effort put into a poem counts for more than the intensity of the emotion felt while writing the poem.
   c. argues that the purpose of poetry is for the reader to feel an emotional connection with the author.
   d. argues that good poetry is produced by poets who think long and deeply about their subjects before they write about it.

Below there is a blank in each question. Choose the word or phrase in the answer choice that best fits the meaning of the sentence as a whole.

5. Before she put a down payment on the house, the would-be buyer had to make sure the house was properly _____ first.
   a. Vacated
   b. Dilapidated
   c. Inspected
   d. Insulated

6. The time had come when Deirdre knew she had to _____ her position at her company in order to go back to school and earn a degree.
   a. Relinquish
   b. Rearrange
   c. Reciprocate
   d. Receive

Questions 7-9 are based on the following passage:

George Washington emerged out of the American Revolution as an unlikely champion of liberty. On June 14, 1775, the Second Continental Congress created the Continental Army, and John Adams, serving in the Congress, nominated Washington to be its first commander. Washington fought under the British during the French and Indian War, and his experience and prestige proved instrumental to the American war effort. Washington provided invaluable leadership, training, and strategy during the Revolutionary War. He emerged from the war as the embodiment of liberty and freedom from tyranny.

After vanquishing the heavily favored British forces, Washington could have pronounced himself as the autocratic leader of the former colonies without any opposition, but he famously refused and returned to his Mount Vernon plantation. His restraint proved his commitment to the fledgling state's republicanism. Washington was later unanimously elected as the first American president. But it is Washington's farewell address that cemented his legacy as a visionary worthy of study.

In 1796, President Washington issued his farewell address by public letter. Washington enlisted his good friend, Alexander Hamilton, in drafting his most famous address. The letter expressed Washington's faith in the Constitution and rule of law. He encouraged his fellow Americans to put aside partisan differences and establish a national union. Washington warned Americans against meddling in foreign affairs and entering military alliances. Additionally, he stated his opposition to national political parties, which he considered partisan and counterproductive.

Americans would be wise to remember Washington's farewell, especially during presidential elections when politics hits a fever pitch. They might want to question the political institutions that were not planned by the Founding Fathers, such as the nomination process and political parties themselves.

7. Which of the following statements is logically based on the information contained in the passage above?
    a. George Washington's background as a wealthy landholder directly led to his faith in equality, liberty, and democracy.
    b. George Washington would have opposed America's involvement in the Second World War.
    c. George Washington would not have been able to write as great a farewell address without the assistance of Alexander Hamilton.
    d. George Washington would probably not approve of modern political parties.

8. Which of the following statements is the best description of the author's purpose in writing this passage about George Washington?
    a. To inform American voters about a Founding Father's sage advice on a contemporary issue and explain its applicability to modern times
    b. To introduce George Washington to readers as a historical figure worthy of study
    c. To note that George Washington was more than a famous military hero
    d. To convince readers that George Washington is a hero of republicanism and liberty

9. In which of the following materials would the author be the most likely to include this passage?
    a. A history textbook
    b. An obituary
    c. A fictional story
    d. A newspaper editorial

The next article is for questions 10-14:

The Old Man and His Grandson

There was once a very old man, whose eyes had become dim,

his ears dull of hearing, his knees trembled, and when he sat at

table he could hardly hold the spoon, and spilt the broth upon

the table-cloth or let it run out of his mouth. His son and his

son's wife were disgusted at this, so the old grandfather at last

had to sit in the corner behind the stove, and they gave him his

food in an earthenware bowl, and not even enough of it. And he

used to look towards the table with his eyes full of tears. Once,

too, his trembling hands could not hold the bowl, and it fell to

the ground and broke. The young wife scolded him, but he said

nothing and only sighed. Then they brought him a wooden

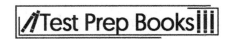

bowl for a few half-pence, out of which he had to eat.

They were once sitting thus when the little grandson of four

years old began to gather together some bits of wood upon the

ground. 'What are you doing there?' asked the father. 'I am

making a little trough,' answered the child, 'for father and

mother to eat out of when I am big.'

The man and his wife looked at each other for a while, and

presently began to cry. Then they took the old grandfather to

the table, and henceforth always let him eat with them, and

likewise said nothing if he did spill a little of anything.
(*Grimms' Fairy Tales*, p. 111)

10. Which of the following most accurately represents the theme of the passage?
    a. Respect your elders
    b. Children will follow their parents' example
    c. You reap what you sow
    d. Loyalty will save your life

11. How is the content in this selection organized?
    a. Chronologically
    b. Problem and solution
    c. Compare and contrast
    d. Order of importance

12. Which character trait most accurately reflects the son and his wife in this story?
    a. Compassion
    b. Understanding
    c. Cruelty
    d. Impatience

13. Where does the story take place?
    a. In the countryside
    b. In America
    c. In a house
    d. In a forest

14. Why do the son and his wife decide to let the old man sit at the table?
    a. Because they felt sorry for him
    b. Because their son told them to
    c. Because the old man would not stop crying
    d. Because they saw their own actions in their son

15. Read the following poem. Which option best expresses the symbolic meaning of the "road" and the overall theme?

> Two roads diverged in a yellow wood,
> And sorry I could not travel both
> And be one traveler, long I stood
> And looked down one as far as I could
> To where it bent in the undergrowth;
> Then took the other, as just as fair,
> And having perhaps the better claim,
> Because it was grassy and wanted wear;
> Though as for that the passing there
> Had worn them really about the same,
> And both that morning equally lay
> In leaves no step had trodden black.
> Oh, I kept the first for another day!
> Yet knowing how way leads on to way,
> I doubted if I should ever come back.
> I shall be telling this with a sigh
> Somewhere ages and ages hence:
> Two roads diverged in a wood, and I—
> I took the one less traveled by,
> And that has made all the difference—Robert Frost, "The Road Not Taken"

a. A divergent spot where the traveler had to choose the correct path to his destination
b. A choice between good and evil that the traveler needs to make
c. The traveler's struggle between his lost love and his future prospects
d. Life's journey and the choices with which humans are faced

16. Which option best exemplifies an author's use of alliteration and personification?
a. Her mood hung about her like a weary cape, very dull from wear.
b. It shuddered, swayed, shook, and screamed its way into dust under hot flames.
c. The house was a starch sentry, warning visitors away.
d. At its shoreline, visitors swore they heard the siren call of the cliffs above.

17. In 1889, Jerome K. Jerome wrote a humorous account of a boating holiday. Originally intended as a chapter in a serious travel guide, the work became a prime example of a comic novel. Read the passage below, noting the word/words in italics. Answer the question that follows.

> I felt rather hurt about this at first; it seemed somehow to be a sort of slight. Why hadn't I got housemaid's knee? Why this invidious reservation? After a while, however, less grasping feelings prevailed. I reflected that I had every other known malady in the pharmacology, and I grew less selfish, and determined to do without housemaid's knee. Gout, in its most malignant stage, it would appear, had seized me without my being aware of it; and *zymosis* I had evidently been

suffering with from boyhood. There were no more diseases after *zymosis*, so I concluded there was nothing else the matter with me.—Jerome K. Jerome, *Three Men in a Boat*

Which definition best fits the word *zymosis*?
- a. Discontent
- b. An infectious disease
- c. Poverty
- d. Bad luck

18. What is the meaning of the word *rookeries* in the following text?

To-day, the plume hunters who do not dare to raid the guarded rookeries are trying to study out the lines of flight of the birds, to and from their feeding-grounds, and shoot them in transit.

- a. Houses in a slum area
- b. A place where hunters gather to trade tools
- c. A place where wardens go to trade stories
- d. A colony of breeding birds

Questions 19-24 are based upon the following passage:

This excerpt is an adaptation from Charles Dickens' speech in Birmingham in England on December 30, 1853 on behalf of the Birmingham and Midland Institute.

My Good Friends,—When I first imparted to the committee of the projected Institute my particular wish that on one of the evenings of my readings here the main body of my audience should be composed of working men and their families, I was animated by two desires; first, by the wish to have the great pleasure of meeting you face to face at this Christmas time, and accompany you myself through one of my little Christmas books; and second, by the wish to have an opportunity of stating publicly in your presence, and in the presence of the committee, my earnest hope that the Institute will, from the beginning, recognise one great principle—strong in reason and justice—which I believe to be essential to the very life of such an Institution. It is, that the working man shall, from the first unto the last, have a share in the management of an Institution which is designed for his benefit, and which calls itself by his name.

I have no fear here of being misunderstood—of being supposed to mean too much in this. If there ever was a time when any one class could of itself do much for its own good, and for the welfare of society—which I greatly doubt—that time is unquestionably past. It is in the fusion of different classes, without confusion; in the bringing together of employers and employed; in the creating of a better common understanding among those whose interests are identical, who depend upon each other, who are vitally essential to each other, and who never can be in unnatural antagonism without deplorable results, that one of the chief principles of a Mechanics' Institution should consist. In this world a great deal of the bitterness among us arises from an imperfect understanding of one another. Erect in Birmingham a great Educational Institution, properly educational; educational of the feelings as well as of the reason; to which all orders of Birmingham men contribute; in which all orders of Birmingham men meet; wherein all orders of Birmingham men are faithfully represented—and you will erect a Temple of Concord here which will be a model edifice to the whole of England.

Contemplating as I do the existence of the Artisans' Committee, which not long ago considered the establishment of the Institute so sensibly, and supported it so heartily, I earnestly entreat the gentlemen—earnest I know in the good work, and who are now among us,—by all means to avoid the great shortcoming of similar institutions; and in asking the working man for his confidence, to set him the great example and give him theirs in return. You will judge for yourselves if I promise too much for the working man, when I say that he will stand by such an enterprise with the utmost of his patience, his perseverance, sense, and support; that I am sure he will need no charitable aid or condescending patronage; but will readily and cheerfully pay for the advantages which it confers; that he will prepare himself in individual cases where he feels that the adverse circumstances around him have rendered it necessary; in a word, that he will feel his responsibility like an honest man, and will most honestly and manfully discharge it. I now proceed to the pleasant task to which I assure you I have looked forward for a long time.

19. Which word is most closely synonymous with the word *patronage* as it appears in the following statement?

      . . . that I am sure he will need no charitable aid or condescending patronage

    a. Auspices
    b. Aberration
    c. Acerbic
    d. Adulation

20. Which term is most closely aligned with the definition of the term *working man* as it is defined in the following passage?

      You will judge for yourselves if I promise too much for the working man, when I say that he will stand by such an enterprise with the utmost of his patience, his perseverance, sense, and support . . .

    a. Plebeian
    b. Viscount
    c. Entrepreneur
    d. Bourgeois

21. Which of the following statements most closely correlates with the definition of the term *working man* as it is defined in the previous question?

    a. A working man is not someone who works for institutions or corporations, but someone who is well versed in the workings of the soul.
    b. A working man is someone who is probably not involved in social activities because the physical demand for work is too high.
    c. A working man is someone who works for wages among the middle class.
    d. The working man has historically taken to the field, to the factory, and now to the screen.

22. Based upon the contextual evidence provided in the passage above, what is the meaning of the term *enterprise* in the third paragraph?
    a. Company
    b. Courage
    c. Game
    d. Cause

23. The speaker addresses his audience as *My Good Friends*—what kind of credibility does this salutation give to the speaker?
    a. The speaker is an employer addressing his employees, so the salutation is a way for the boss to bridge the gap between himself and his employees.
    b. The speaker's salutation is one from an entertainer to his audience and uses the friendly language to connect to his audience before a serious speech.
    c. The salutation gives the serious speech that follows a somber tone, as it is used ironically.
    d. The speech is one from a politician to the public, so the salutation is used to grab the audience's attention.

24. According to the aforementioned passage, what is the speaker's second desire for his time in front of the audience?
    a. To read a Christmas story
    b. For the working man to have a say in his institution which is designed for his benefit
    c. To have an opportunity to stand in their presence
    d. For the life of the institution to be essential to the audience as a whole

Writing

Directions for questions 1 and 2

Select the best version of the underlined part of the sentence. The first choice is the same as the original sentence. If you think the original sentence is best, choose the first answer.

1. It is necessary for instructors to offer tutoring <u>to any students who need extra help in the class.</u>
    a. to any students who need extra help in the class.
    b. for any students that need extra help in the class.
    c. with any students who need extra help in the class.
    d. for any students needing any extra help in their class.

2. <u>Because many people</u> feel there are too many distractions to get any work done, I actually enjoy working from home.
    a. Because many people
    b. While many people
    c. Maybe many people
    d. With most people

Read the following section about Fred Hampton and answer Questions 3-15.

    1 Fred Hampton desired to see lasting social change for African American people through nonviolent means and community recognition. (3) <u>In the meantime,</u> he became an African

American activist during the American Civil Rights Movement and led the Chicago chapter of the Black Panther Party.

Hampton's Education

2 Hampton was born and raised (4) <u>in Maywood of Chicago, Illinois in 1948.</u> Gifted academically and a natural athlete, he became a stellar baseball player in high school. (5) <u>After graduating from Proviso East High School in 1966, he later went on to study law at Triton Junior College. While studying at Triton, Hampton joined and became a leader of the National Association for the Advancement of Colored People (NAACP). As a result of his leadership, the NAACP gained more than 500 members.</u> Hampton worked relentlessly to acquire recreational facilities in the neighborhood and improve the educational resources provided to the impoverished black community of Maywood.

The Black Panthers

3 The Black Panther Party (BPP) (6) <u>was another that </u>formed around the same time as and was similar in function to the NAACP. Hampton was quickly attracted to the (7) <u>Black Panther Party's approach</u> to the fight for equal rights for African Americans. Hampton eventually joined the chapter and relocated to downtown Chicago to be closer to its headquarters.

4 His charismatic personality, organizational abilities, sheer determination, and rhetorical skills (8) <u>enable him to quickly rise</u> through the chapter's ranks. Hampton soon became the leader of the Chicago chapter of the BPP where he organized rallies, taught political education classes, and established a free medical clinic. (9) <u>He also took part in the community police supervision project. He played an instrumental role</u> in the BPP breakfast program for impoverished African American children.

5 Hampton's (10) <u>greatest acheivement</u> as the <u>leader</u> of the BPP may be his fight against street gang violence in Chicago. In 1969, (11) <u>Hampton was held by a press conference</u> where he made the gangs agree to a nonaggression pact known as the Rainbow Coalition. As a result of the pact, a multiracial alliance between blacks, Puerto Ricans, and poor youth was developed.

Assassination

6 (12) <u>As the Black Panther Party's popularity and influence grew, the Federal Bureau of Investigation (FBI) placed the group under constant surveillance.</u> In an attempt to neutralize the party, the FBI launched several harassment campaigns against the BPP, raided its headquarters in Chicago three times, and arrested over one hundred of the group's members. Hampton was shot during such a raid that occurred on the morning of December 4th 1969.

7 (13) <u>In 1976; seven years after the event,</u> it was revealed that William O'Neal, Hampton's trusted bodyguard, was an undercover FBI agent. (14) <u>O'Neal will provide</u> the FBI with detailed floor plans of the BPP's headquarters, identifying the exact location of Hampton's bed. It was because of these floor plans that the police were able to target and kill Hampton.

8 The assassination of Hampton fueled outrage amongst the African American community. It was not until years after the assassination that the police admitted wrongdoing. (15) <u>The Chicago City Council now are commemorating December 4th as Fred Hampton Day.</u>

3. In context, which is the best version of the underlined portion of this sentence (reproduced below)?

In the meantime, he became an African American activist during the American Civil Rights Movement and led the Chicago chapter of the Black Panther Party.

a. (as it is now)
b. Unfortunately,
c. Finally,
d. As a result,

4. Which is the best version of the underlined portion of this sentence (reproduced below)?

Hampton was born and raised in Maywood of Chicago, Illinois in 1948.

a. (as it is now)
b. in Maywood, of Chicago, Illinois in 1948.
c. in Maywood of Chicago, Illinois, in 1948.
d. in Chicago, Illinois of Maywood in 1948.

5. Which of the following sentences, if any, should begin a new paragraph?

After graduating from Proviso East High School in 1966, he later went on to study law at Triton Junior College. While studying at Triton, Hampton joined and became a leader of the National Association for the Advancement of Colored People (NAACP). As a result of his leadership, the NAACP gained more than 500 members.

a. There should be no new paragraph.
b. After graduating from Proviso East High School in 1966, he later went on to study law at Triton Junior College.
c. While studying at Triton, Hampton joined and became a leader of the National Association for the Advancement of Colored People (NAACP).
d. As a result of his leadership, the NAACP gained more than 500 members.

6. Which of the following facts would be the most relevant to include here?

The Black Panther Party (BPP) was another that formed around the same time as and was similar in function to the NAACP.

a. best as written
b. was another activist group that
c. had a lot of members that
d. was another school that

115

7. Which is the best version of the underlined portion of this sentence (reproduced below)?

Hampton was quickly attracted to the <u>Black Panther Party's approach</u> to the fight for equal rights for African Americans.

a. (as it is now)
b. Black Panther Parties approach
c. Black Panther Partys' approach
d. Black Panther Parties' approach

8. Which is the best version of the underlined portion of this sentence (reproduced below)?

His charismatic personality, organizational abilities, sheer determination, and rhetorical skills <u>enable him to quickly rise</u> through the chapter's ranks.

a. (as it is now)
b. are enabling him to quickly rise
c. enabled him to quickly rise
d. will enable him to quickly rise

9. Which is the best version of the underlined portion of this sentence (reproduced below)?

<u>He also took part in the community police supervision project. He played an instrumental role</u> in the BPP breakfast program for impoverished African American children.

a. (as it is now)
b. He also took part in the community police supervision project but played an instrumental role
c. He also took part in the community police supervision project, he played an instrumental role
d. He also took part in the community police supervision project and played an instrumental role

10. Which of these, if any, is misspelled?

Hampton's <u>greatest</u> <u>acheivement</u> as the <u>leader</u> of the BPP may be his fight against street gang violence in Chicago.

a. None of these are misspelled.
b. greatest
c. acheivement
d. leader

11. Which is the best version of the underlined portion of this sentence (reproduced below)?

In 1969, <u>Hampton was held by a press conference</u> where he made the gangs agree to a nonaggression pact known as the Rainbow Coalition.

a. (as it is now)
b. Hampton held a press conference
c. Hampton, holding a press conference
d. Hampton to hold a press conference

12. Which is the best version of the underlined portion of this sentence (reproduced below)?

As the Black Panther Party's popularity and influence grew, the Federal Bureau of Investigation (FBI) placed the group under constant surveillance.

a. (as it is now)
b. The Federal Bureau of Investigation (FBI) placed the group under constant surveillance as the Black Panther Party's popularity and influence grew.
c. Placing the group under constant surveillance, the Black Panther Party's popularity and influence grew.
d. As their influence and popularity grew, the FBI placed the group under constant surveillance.

13. Which is the best version of the underlined portion of this sentence (reproduced below)?

In 1976; seven years after the event, it was revealed that William O'Neal, Hampton's trusted bodyguard, was an undercover FBI agent.

a. (as it is now)
b. In 1976, seven years after the event,
c. In 1976 seven years after the event,
d. In 1976. Seven years after the event,

14. Which is the best version of the underlined portion of this sentence (reproduced below)?

O'Neal will provide the FBI with detailed floor plans of the BPP's headquarters, identifying the exact location of Hampton's bed.

a. (as it is now)
b. O'Neal provides
c. O'Neal provided
d. O'Neal, providing

15. Which is the best version of the underlined portion of this sentence (reproduced below)?

The Chicago City Council now are commemorating December 4th as Fred Hampton Day.

a. (as it is now)
b. Fred Hampton Day by the Chicago City Council, December 4, is now commemorated.
c. Now commemorated December 4th is Fred Hampton Day.
d. The Chicago City Council now commemorates December 4th as Fred Hampton Day.

16. Which of the following sentences uses correct punctuation?
    a. Carole is not currently working; her focus is on her children at the moment.
    b. Carole is not currently working and her focus is on her children at the moment.
    c. Carole is not currently working, her focus is on her children at the moment.
    d. Carole is not currently working her focus is on her children at the moment.

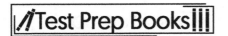

17. Which of these examples shows incorrect use of subject-verb agreement?
    a. Neither of the cars are parked on the street.
    b. Both of my kids are going to camp this summer.
    c. Any of your friends are welcome to join us on the trip in November.
    d. Each of the clothing options is appropriate for the job interview.

18. When it gets warm in the spring, _____ and _____ like to go fishing at Cobbs Creek.
Which of the following word pairs should be used in the blanks above?
    a. me, him
    b. he, I
    c. him, I
    d. he, me

19. Which of the following examples uses correct punctuation?
    a. Recommended supplies for the hunting trip include the following: rain gear, large backpack, hiking boots, flashlight, and non-perishable foods.
    b. I left the store, because I forgot my wallet.
    c. As soon as the team checked into the hotel; they met in the lobby for a group photo.
    d. None of the furniture came in on time: so they weren't able to move in to the new apartment.

20. Which of the following sentences uses correct subject-verb agreement?
    a. There is two constellations that can be seen from the back of the house.
    b. At least four of the sheep needs to be sheared before the end of summer.
    c. Lots of people were auditioning for the singing competition on Saturday.
    d. Everyone in the group have completed the assignment on time.

Essay

Please read the prompt below and answer in an essay format in 300–600 words.

Coaches of kids' sports teams are increasingly concerned about the behavior of parents at games. Parents are screaming and cursing at coaches, officials, players, and other parents. Physical fights have even broken out at games. Parents need to be reminded that coaches are volunteers, giving up their time and energy to help kids develop in their chosen sport. The goal of kids' sports teams is to learn and develop skills, but it's also to have fun. When parents are out of control at games and practices, it takes the fun out of the sport.

1. Analyze and evaluate the passage given.

2. State and develop your own perspective.

3. Explain the relationship between your perspective and the one given.

Answer Explanations for Practice Test #2

Math

1. C: The numbers needed are those that add to -2 and multiply to -8. The difference between 2 and 4 is 2. Their product is 8, and -4 and 2 will work. Therefore, $x^2 - 2x - 8 = (x - 4)(x + 2)$. The latter has roots 4 and -2 or 1 ± 3. Alternatively, you can use the quadratic formula and then simplify.

2. A: Parallel lines have the same slope. The slope of Choice C can be seen to be 1/3 by dividing both sides by 3. The others are in standard form $Ax + By = C$, for which the slope is given by $\frac{-A}{B}$. The slope of A is 3, the slope of B is 4, and the slope of D is 1.

3. C: We are trying to find x, the number of red cans. The equation can be set up like this:

$$x + 2(10 - x) = 16$$

The left x is actually multiplied by $1, the price per red can. Since we know Jessica bought 10 total cans, $10 - x$ is the number blue cans that she bought. We multiply the number of blue cans by $2, the price per blue can.

That should all equal $16, the total amount of money that Jessica spent. Working that out gives us:

$$x + 20 - 2x = 16$$

$$20 - x = 16$$

$$x = 4$$

4. A: If each man gains 10 pounds, every original data point will increase by 10 pounds. Therefore, the man with the original median will still have the median value, but that value will increase by 10. The smallest value and largest value will also increase by 10 and, therefore, the difference between the two won't change. The range does not change in value and, thus, remains the same.

5. A: Check each value, but it is easiest to use the quadratic formula, which gives:

$$x = \frac{2 \pm \sqrt{(-2)^2 - 4(1)(-2)}}{2} = 1 \pm \frac{\sqrt{12}}{2} = 1 \pm \frac{2\sqrt{3}}{2} = 1 \pm \sqrt{3}$$

The only one of these which appears as an answer choice is $1 + \sqrt{3}$.

6. A: Finding the product means distributing one polynomial over the other so that each term in the first is multiplied by each term in the second. Then, like terms can be collected. Multiplying the factors yields the expression:

$$20x^3 + 4x^2 + 24x - 40x^2 - 8x - 48$$

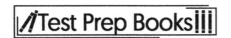
Collecting like terms means adding the x^2 terms and adding the x terms. The final answer after simplifying the expression is:

$$20x^3 - 36x^2 + 16x - 48$$

7. B: The car is traveling at a speed of five meters per second. On the interval from one to three seconds, the position changes by ten meters. By making this change in position over time into a rate, the speed becomes ten meters in two seconds or five meters in one second.

8. B: The y-intercept of an equation is found where the x-value is zero. Plugging zero into the equation for x, the first two terms cancel out, leaving -4.

9. C: To solve for the value of b, both sides of the equation need to be equalized.

Start by cancelling out the lower value of -4 by adding 4 to both sides:

$$5b - 4 = 2b + 17$$

$$5b = 2b + 21$$

The variable b is the same on each side, so subtract the lower 2b from each side:

$$5b = 2b + 21$$

$$3b = 21$$

Then divide both sides by 3 to get the value of b:

$$\frac{3b}{3} = \frac{21}{3}$$

$$b = 7$$

10. D: To find the average of a set of values, add the values together and then divide by the total number of values. In this case, include the unknown value of what Dwayne needs to score on his next test, in order to solve it.

$$\frac{78 + 92 + 83 + 97 + x}{5} = 90$$

Add the unknown value to the new average total, which is 5. Then multiply each side by 5 to simplify the equation, resulting in:

$$78 + 92 + 83 + 97 + x = 450$$

$$350 + x = 450$$

$$x = 100$$

Dwayne would need to get a perfect score of 100 in order to get an average of at least 90.

Test this answer by substituting back into the original formula.

$$\frac{78 + 92 + 83 + 97 + 100}{5} = 90$$

11. A: To calculate the range in a set of data, subtract the lowest value from the highest value. In this graph, the range of Mr. Lennon's students is 5, which can be seen physically in the graph as having the smallest difference between the highest value and the lowest value compared with the other teachers.

12. D: The addition rule is necessary to determine the probability because a 6 can be rolled on either roll of the die. The rule used is $P(A \text{ or } B) = P(A) + P(B) - P(A \text{ and } B)$. The probability of a 6 being individually rolled is $^1/_6$ and the probability of a 6 being rolled twice is $\frac{1}{6} \times \frac{1}{6} = \frac{1}{36}$. Therefore, the probability that a 6 is rolled at least once is $\frac{1}{6} + \frac{1}{6} - \frac{1}{36} = \frac{11}{36}$.

13. D: When an ordered pair is reflected over an axis, the sign of one of the coordinates must change. When it's reflected over the x-axis, the sign of the y-coordinate must change. The x-value remains the same. Therefore, the new ordered pair is $(-3, 4)$.

14. D: $x \leq -5$. When solving a linear equation or inequality:

Distribution is performed if necessary: $-3(x + 4) \rightarrow -3x - 12 \geq x + 8$. This means that any like terms on the same side of the equation/inequality are combined.

The equation/inequality is manipulated to get the variable on one side. In this case, subtracting x from both sides produces $-4x - 12 \geq 8$.

The variable is isolated using inverse operations to undo addition/subtraction. Adding 12 to both sides produces $-4x \geq 20$.

The variable is isolated using inverse operations to undo multiplication/division. Remember if dividing by a negative number, the relationship of the inequality reverses, so the sign is flipped. In this case, dividing by -4 on both sides produces $x \leq -5$.

15. C: $y = 40x + 300$. In this scenario, the variables are the number of sales and Karen's weekly pay. The weekly pay depends on the number of sales. Therefore, weekly pay is the dependent variable (y), and the number of sales is the independent variable (x). Each pair of values from the table can be written as an ordered pair (x, y): (2, 380), (7, 580), (4, 460), (8, 620). The ordered pairs can be substituted into the equations to see which creates true statements (both sides equal) for each pair. Even if one ordered pair produces equal values for a given equation, the other three ordered pairs must be checked. The only equation which is true for all four ordered pairs is $y = 40x + 300$:

$$380 = 40(2) + 300 \rightarrow 380 = 380$$

$$580 = 40(7) + 300 \rightarrow 580 = 580$$

$$460 = 40(4) + 300 \rightarrow 460 = 460$$

$$620 = 40(8) + 300 \rightarrow 620 = 620$$

16. B: For an ordered pair to be a solution to a system of inequalities, it must make a true statement for BOTH inequalities when substituting its values for x and y. Substituting $(-3,-2)$ into the inequalities produces $(-2) > 2(-3) - 3 \rightarrow -2 > -9$ and $(-2) < -4(-3) + 8 \rightarrow -2 < 20$. Both are true statements.

17. C: 216cm. Because area is a two-dimensional measurement, the dimensions are multiplied by a scale that is squared to determine the scale of the corresponding areas. The dimensions of the rectangle are multiplied by a scale of 3. Therefore, the area is multiplied by a scale of 3^2 (which is equal to 9): $24cm \times 9 = 216cm$.

18. B: The figure is composed of three sides of a square and a semicircle. The sides of the square are simply added: 8 + 8 + 8 = 24 inches. The circumference of a circle is found by the equation C = 2πr. The radius is 4 in, so the circumference of the circle is 25.13 in. Only half of the circle makes up the outer border of the figure (part of the perimeter) so half of 25.13 in is 12.565 in. Therefore, the total perimeter is: 24 in + 12.565 in = 36.565 in. The other answer choices use the incorrect formula or fail to include all of the necessary sides.

19. C: An x-intercept is the point where the graph crosses the x-axis. At this point, the value of y is 0. To determine if an equation has an x-intercept of -2, substitute -2 for x, and calculate the value of y. If the value of -2 for x corresponds with a y-value of 0, then the equation has an x-intercept of -2. The only answer choice that produces this result is Choice C:

$$0 = (-2)^2 + 5(-2) + 6$$

20. D: Three girls for every two boys can be expressed as a ratio: 3:2. This can be visualized as splitting the school into 5 groups: 3 girl groups and 2 boy groups. The number of students which are in each group can be found by dividing the total number of students by 5:

$$\frac{650 \text{ students}}{5 \text{ groups}} = \frac{130 \text{ students}}{\text{group}}$$

To find the total number of girls, multiply the number of students per group (130) by the number of girl groups in the school (3). This equals 390, Choice D.

Reading

1. D: Passage A argues that the source of great poetry comes from emotions within, while passage B argues that great poetry is based off the skill and expertise of the poet. Choice A is incorrect; both passages attempt to explain how great poetry is created, not its aim or its results. The "purpose" passage A talks about are the author's individual poems, but the main idea of the passage is contained in the middle, starting with "For all good poetry." Choice B is incorrect; neither passages make any mention of nature or urban life. Choice C is incorrect; passage B does use epic poetry for most of its examples, but it is not solely concerned with epic poetry. It recalls Keats' "Ode to a Nightingale" at the very end. This is not the best answer choice.

2. A: The author is explaining that the feelings portrayed in Keats' ode are brought forth and signified by the symbolism of the nightingale used within the language of the poem. Eliot is relating it back to the original point of the passage: that great poetry doesn't simply come from powerful emotion, but that it is that emotion coupled with the literary devices of the poet that makes great poetry. Some of the other choices have similar wording to the last sentence, but they are not the best answer choice.

3. C: That feelings from the past will accumulate into representations of thought. The poet is then responsible for putting these direct thoughts onto paper, creating a poem. Therefore, the original source of poetry is from the past feelings of the poet. This is explained by the author beginning with the sentence "For our continued influxes of feeling . . . " The author is very clear about feelings from the past accumulating into representations of thought, so all the other answer choices are incorrect.

4. B: The author of passage B argues that the effort put into a poem counts for more than the intensity of the emotion felt while writing the poem. Choice *A* is the opposite of what passage B argues. Choices *C* and *D* are more closely related to the ideas of passage A.

5. C: The best word here is inspected, which means *approved* or *investigated*. Choice *A*, vacated, means to leave empty. While this is a possible choice, it is not the best choice because it is assumed that the house would be vacated before it was bought. Choice *B*, dilapidated, means destroyed. Someone buying a house does not want the house to be left destroyed, so this is incorrect. Choice *D, insulated*, means that the house would be properly enclosed against the loss of heat or the intrusion of sound. Although this is a possible answer, this would be included in an inspection. Therefore, Choice *D* is incorrect.

6. A: Relinquish. Relinquish most closely means to give up or let go. Deirdre knew she had to give up her position in order to go back to school. Choice *B*, rearrange, means to shift or change. Choice *C*, reciprocate, means to exchange or swap. Choice *D*, receive, means to accept.

7. D: Although Washington was from a wealthy background, the passage does not say that his wealth led to his republican ideals, so Choice *A* is not supported. Choice *B* also does not follow from the passage. Washington's warning against meddling in foreign affairs does not mean that he would oppose wars of every kind, so Choice *B* is wrong. Choice *C* is also unjustified since the author does not indicate that Alexander Hamilton's assistance was absolutely necessary. Choice *D* is correct because the farewell address clearly opposes political parties and partisanship. The author then notes that presidential elections often hit a fever pitch of partisanship. Thus, it follows that George Washington would not approve of modern political parties and their involvement in presidential elections.

8. A: The author finishes the passage by applying Washington's farewell address to modern politics, so the purpose probably includes this application. Choice *B* is wrong because George Washington is already a well-established historical figure; furthermore, the passage does not seek to introduce him. Choice *C* is wrong because the author is not fighting a common perception that Washington was merely a military hero. Choice *D* is wrong because the author is not convincing readers. Persuasion does not correspond to the passage. Choice *A* states the primary purpose.

9. D: Choice *A* is wrong because the last paragraph is not appropriate for a history textbook. Choice *B* is false because the piece is not a notice or announcement of Washington's death. Choice *C* is clearly false because it is not fiction, but a piece of historical writing. Choice *D* is correct. The passage is most likely to appear in a newspaper editorial because it cites information relevant and applicable to the present day, which is a popular format in editorials.

10. B: Choice *A* is incorrect because it does not fit with the primary purpose of this passage, which is to tell a story of how a child plans to treat his parents when he sees the way they treat his grandfather. It is trying to remind readers to treat others with respect because that is how one wants to be treated, and that this does not apply only to elderly people. Choice *B* fits most appropriately with the primary purpose, since the son and wife see that they will be treated unfairly because they witness that their child plans to do it to them when they are older. To "reap what you sow" means that there are repercussions for every action. This may seem like the correct answer; however, the parents do not actually have to eat out of a trough later in life. They don't actually experience any repercussions. Even though it may be argued that the boy is being loyal to his grandfather, this does not fit with the primary purpose. The boy also never mentions that his actions are because he cares for his grandfather; rather, he simply mirrors the behaviors of his parents.

11. A: Choice *A* is correct because it follows a series of events that happen in order, one right after the other. First the grandfather spills his food, then his son puts him in a corner, then the child makes a trough for his parents to eat out of when he's older, and finally the parents welcome the old man back to the table. Choice *B* is incorrect as even though it could be argued that the way they treat the old man is a problem, there really isn't a solution to the problem, even though they stop treating him badly. Also, problem and solution styles generally do not follow a chronological timeline. Choice *C* is incorrect because events in the passage are not compared and contrasted; this is not a primary organizational structure of the passage. Choice *D* is incorrect because there is no language to indicate that one person or event is more important than the other.

12. C: Although they do show him compassion in the end, it is not because they feel compassionate for him, but instead, it is because they recognize that their son plans to treat them the way they are treating the old man when they are older. So, they treat the old man the way they would want to be treated. Understanding is not the overall attitude they feel toward the old man, and it is only in realizing the cruelty of their behavior that they understand how they have been treating him. Choice *C* is correct because it condenses the actions of the son and his wife into a single word. Refusing to let the old man sit at the table when he clearly needs help and looks at the table with tear-filled eyes is a cruel thing to do. Choice *D* may be tempting to pick as they *are* impatient with him, but it's not the best answer. People can be impatient without being cruel.

13. C: Choice *A* is incorrect as there is no descriptive language to indicate that they are in the countryside. Choice *B* is incorrect because the passage has no language or descriptions to indicate they are in America. Choice *C* is correct because the setting contains elements of a house: a table, a stove, and a corner. Choice *D* may be tempting as there is mention of "bits of wood upon the ground," but as there are no other elements of a forest in the story, this is not the correct answer.

14. D: The parents allow the old man to sit at the table because their son starts to make them a trough, so their motivation in letting him eat at the table is not because they feel sorry for him, but because they don't want their son to treat them that way when they are old. This makes Choice *A* incorrect. Their son did not tell them to let the old man sit at the table, so Choice *B* is incorrect. In the story, it mentions that even after the old man has eyes full of tears, the wife gave him a cheap wooden bowl to eat out of, so clearly his crying did not make them stop treating him badly, making Choice *C* incorrect. Choice *D* is correct because the parents let the old man sit at the table as a result of the boy mimicking their behavior.

15. D: Choice *D* correctly summarizes Frost's theme of life's journey and the choices one makes. While Choice *A* can be seen as an interpretation, it is a literal one and is incorrect. Literal is not symbolic. Choice *B* presents the idea of good and evil as a theme, and the poem does not specify this struggle for the traveler. Choice *C* is a similarly incorrect answer. Love is not the theme.

16. B: Only Choice *B* uses both repetitive beginning sounds (alliteration) and personification—the portrayal of a building as a human crumbling under a fire. Choice *A* is a simile and does not utilize alliteration or the use of consistent consonant sounds for effect. Choice *C* is a metaphor and does not utilize alliteration. Choice *D* describes neither alliteration nor personification.

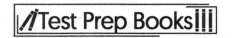

17. B: The correct answer is an infectious disease. By reading context, all other options can be eliminated since the author restates zymosis as disease.

18. D: A *rookery* is a colony of breeding birds. Although *rookery* could mean Choice *A*, houses in a slum area, it does not make sense in this context. Choices *B* and *C* are both incorrect, as this is not a place for hunters to trade tools or for wardens to trade stories.

19. A: The word *patronage* most nearly means *auspices*, which means *protection* or *support*. Choice *B*, *aberration*, means *deformity* and does not make sense within the context of the sentence. Choice *C*, *acerbic,* means *bitter* and also does not make sense in the sentence. Choice *D*, *adulation*, is a positive word meaning *praise*, and thus does not fit with the word *condescending* in the sentence.

20. D: *Working man* is most closely aligned with Choice *D*, *bourgeois.* In the context of the speech, the word *bourgeois* means *working* or *middle class*. Choice *A*, *Plebeian*, does suggest *common people*; however, this is a term that is specific to ancient Rome. Choice *B*, *viscount*, is a European title used to describe a specific degree of nobility. Choice *C*, *entrepreneur*, is a person who operates their own business.

21. C: In the context of the speech, the term *working man* most closely correlates with Choice *C*, *working man is someone who works for wages among the middle class.* Choice *A* is not mentioned in the passage and is off-topic. Choice *B* may be true in some cases, but it does not reflect the sentiment described for the term *working man* in the passage. Choice *D* may also be arguably true. However, it is not given as a definition but as *acts* of the working man, and the topics of *field, factory,* and *screen* are not mentioned in the passage.

22. D: *Enterprise* most closely means *cause.* Choices *A, B,* and *C* are all related to the term *enterprise.* However, Dickens speaks of a *cause* here, not a company, courage, or a game. *He will stand by such an enterprise* is a call to stand by a cause to enable the working man to have a certain autonomy over his own economic standing. The very first paragraph ends with the statement that the working man *shall . . . have a share in the management of an institution which is designed for his benefit.*

23. B: The speaker's salutation is one from an entertainer to his audience and uses the friendly language to connect to his audience before a serious speech. Recall in the first paragraph that the speaker is there to "accompany [the audience] . . . through one of my little Christmas books," making him an author there to entertain the crowd with his own writing. The speech preceding the reading is the passage itself, and, as the tone indicates, a serious speech addressing the "working man." Although the passage speaks of employers and employees, the speaker himself is not an employer of the audience, so Choice *A* is incorrect. Choice *C* is also incorrect, as the salutation is not used ironically, but sincerely, as the speech addresses the wellbeing of the crowd. Choice *D* is incorrect because the speech is not given by a politician, but by a writer.

24. B: For the working man to have a say in his institution which is designed for his benefit. Choice *A* is incorrect because that is the speaker's *first* desire, not his second. Choices *C* and *D* are tricky because the language of both of these is mentioned after the word *second*. However, the speaker doesn't get to the second wish until the next sentence. Choices *C* and *D* are merely prepositions preparing for the statement of the main clause, Choice *B*.

Writing

1. A: Answer Choice *A* uses the best, most concise word choice. Choice *B* uses the pronoun *that* to refer to people instead of *who*. *Choice C* incorrectly uses the preposition *with*. Choice *D* uses the preposition *for* and the additional word *any*, making the sentence wordy and less clear.

2. B: Choice *B* uses the best choice of words to create a subordinate and independent clause. In Choice *A*, *because* makes it seem like this is the reason I enjoy working from home, which is incorrect. In Choice *C*, the word *maybe* creates two independent clauses, which are not joined properly with a comma. Choice *D* uses *with*, which does not make grammatical sense.

3. D: Choice *D* is correct because Fred Hampton becoming an activist was a direct result of him wanting to see lasting social change for Black people. Choice *A* doesn't make sense because "In the meantime" denotes something happening at the same time as another thing. Choice *B* is incorrect because the text's tone does not indicate that becoming a civil rights activist is an unfortunate path. Choice *C* is incorrect because "Finally" indicates something that comes last in a series of events, and the word in question is at the beginning of the introductory paragraph.

4. A: Choice *A* is correct because there should be a comma between the city and state. Choice *B* is incorrect because the comma after "Maywood" interrupts the phrase "Maywood of Chicago." Choice *C* is incorrect because a comma after "Illinois" is unnecessary. Finally, Choice *D* is incorrect because the order of the sentence designates that Chicago, Illinois is in Maywood, which is incorrect.

5. C: This is a difficult question. The paragraph is incorrect as-is because it is too long and thus loses the reader halfway through. Choice *C* is correct because if the new paragraph began with "While studying at Triton," we would see a smooth transition from one paragraph to the next. We can also see how the two paragraphs are logically split in two. The first half of the paragraph talks about where he studied. The second half of the paragraph talks about the NAACP and the result of his leadership in the association. If we look at the passage as a whole, we can see that there are two main topics that should be broken into two separate paragraphs.

6. B: The BPP "was another activist group that . . ." We can figure out this answer by looking at context clues. We know that the BPP is "similar in function" to the NAACP. To find out what the NAACP's function is, we must look at the previous sentences. We know from above that the NAACP is an activist group, so we can assume that the BPP is also an activist group.

7. A: Choice *A* is correct because the Black Panther Party is one entity; therefore, the possession should show the "Party's approach" with the apostrophe between the "y" and the "s." Choice *B* is incorrect because the word "Parties" should not be plural. Choice *C* is incorrect because the apostrophe indicates that the word "Partys" is plural. The plural of "party" is "parties." Choice *D* is incorrect because, again, the word "parties" should not be plural; instead, it is one unified party.

8. C: Choice *C* is correct because the passage is told in past tense, and "enabled" is a past tense verb. Choice *A*, "enable," is present tense. Choice *B*, "are enabling," is a present participle, which suggests a continuing action. Choice *D*, "will enable," is future tense.

9. D: Choice *D* is correct because the conjunction "and" is the best way to combine the two independent clauses. Choice *A* is incorrect because the word "he" becomes repetitive since the two clauses can be joined together. Choice *B* is incorrect because the conjunction "but" indicates a contrast, and there is no contrast between the two clauses. Choice *C* is incorrect because the introduction of the comma after "project" with no conjunction creates a comma splice.

10. C: The word "acheivement" is misspelled. Remember the rules for "*i* before *e* except after *c*." Choices *B* and *D*, "greatest" and "leader," are both spelled correctly.

11. B: Choice *B* is correct because it provides the correct verb tense and verb form. Choice *A* is incorrect; Hampton was not "held by a press conference"—rather, he held a press conference. The passage indicates that he "made the gangs agree to a nonaggression pact," implying that it was Hampton who was doing the speaking for this conference. Choice *C* is incorrect because, with this use of the sentence, it would create a fragment because the verb "holding" has no helping verb in front of it. Choice *D* is incorrect because it adds an infinitive ("to hold") where a past tense form of a verb should be.

12. A: Choice *A* is correct because it provides the most clarity. Choice *B* is incorrect because it doesn't name the group until the end, so the phrase "the group" is vague. Choice *C* is incorrect because it indicates that the BPP's popularity grew as a result of placing the group under constant surveillance, which is incorrect. Choice *D* is incorrect because there is a misplaced modifier; this sentence actually says that the FBI's influence and popularity grew, which is incorrect.

13. B: Choice *B* is correct. Choice *A* is incorrect because there should be an independent clause on either side of a semicolon, and the phrase "In 1976" is not an independent clause. Choice *C* is incorrect because there should be a comma after introductory phrases in general, such as "In 1976," and Choice *C* omits a comma. Choice *D* is incorrect because the sentence "In 1976." is a fragment.

14. C: Choice *C* is correct because the past tense verb "provided" fits in with the rest of the verb tense throughout the passage. Choice *A*, "will provide," is future tense. Choice *B*, "provides," is present tense. Choice *D*, "providing," is a present participle, which means the action is continuous.

15. D: The correct answer is Choice *D* because this statement provides the most clarity. Choice *A* is incorrect because the noun "Chicago City Council" acts as one, so the verb "are" should be singular, not plural. Choice *B* is incorrect because it is perhaps the most confusingly worded out of all the answer choices; the phrase "December 4" interrupts the sentence without any indication of purpose. Choice *C* is incorrect because it is too vague and leaves out *who* does the commemorating.

16. A: Choice *A* is correctly punctuated because it uses a semicolon to join two independent clauses that are related in meaning. Each of these clauses could function as an independent sentence. Choice *B* is incorrect because the conjunction is not preceded by a comma. A comma and conjunction should be used together to join independent clauses. Choice *C* is incorrect because a comma should only be used to join independent sentences when it also includes a coordinating conjunction such as *and* or *so*. Choice *D* does not use punctuation to join the independent clauses, so it is considered a fused (same as a run-on) sentence.

17. A: Choice *A* uses incorrect subject-verb agreement because the indefinite pronoun *neither* is singular and must use the singular verb form *is*. The pronoun *both* is plural and uses the plural verb form of *are*. The pronoun *any* can be either singular or plural. In this example, it is used as a plural, so the plural verb form *are* is used. The pronoun *each* is singular and uses the singular verb form *is*.

18. B: Choice *B* is correct because the pronouns *he* and *I* are in the subjective case. *He* and *I* are the subjects of the verb *like* in the independent clause of the sentence. Choice *A, C,* and *D* are incorrect because they all contain at least one objective pronoun (*me* and *him*). Objective pronouns should not be used as the subject of the sentence, but rather, they should come as an object of a verb. To test for correct pronoun usage, try reading the pronouns as if they were the only pronoun in the sentence. For example, *he* and *me* may appear to be the correct answer choices, but try reading them as the only pronoun.

He like[s] to go fishing...

Me like to go fishing...

When looked at that way, *me* is an obviously incorrect choice.

19. A: In this example, a colon is correctly used to introduce a series of items. Choice *B* places an unnecessary comma before the word *because*. A comma is not needed before the word *because* when it introduces a dependent clause at the end of a sentence and provides necessary information to understand the sentence. Choice *C* is incorrect because it uses a semi-colon instead of a comma to join a dependent clause and an independent clause. Choice *D* is incorrect because it uses a colon in place of a comma and coordinating conjunction to join two independent clauses.

20. C: The simple subject of this sentence, the word *lots*, is plural. It agrees with the plural verb form *were*. Choice *A* is incorrect, because the simple subject *there*, referring to the two constellations, is considered plural. It does not agree with the singular verb form *is*. In Choice *B*, the singular subject *four*, does not agree with the plural verb form *needs*. In Choice *D* the plural subject *everyone* does not agree with the singular verb form *have*.

TSI Practice Test #3

Math

1. If $6t + 4 = 16$, what is t?
   a. 1
   b. 2
   c. 3
   d. 4

2. The variable y is directly proportional to x. If $y = 3$ when $x = 5$, then what is y when $x = 20$?
   a. 10
   b. 12
   c. 14
   d. 16

3. A line passes through the point (1, 2) and crosses the y-axis at $y = 1$. Which of the following is an equation for this line?
   a. $y = 2x$
   b. $y = x + 1$
   c. $x + y = 1$
   d. $y = \frac{x}{2} - 2$

4. There are $4x + 1$ treats in each party favor bag. If a total of $60x + 15$ treats are distributed, how many bags are given out?
   a. 15
   b. 16
   c. 20
   d. 22

5. Apples cost $2 each, while bananas cost $3 each. Maria purchased 10 fruits in total and spent $22. How many apples did she buy?
   a. 5
   b. 6
   c. 7
   d. 8

6. What are the polynomial roots of $x^2 + x - 2$?
   a. 1 and -2
   b. -1 and 2
   c. 2 and -2
   d. 9 and 13

7. What is the y-intercept of $y = x^{5/3} + (x - 3)(x + 1)$?
   a. 3.5
   b. 7.6
   c. -3
   d. -15.1

8. $x^4 - 16$ can be simplified to which of the following?

    a. $(x^2 - 4)(x^2 + 4)$
    b. $(x^2 + 4)(x^2 + 4)$
    c. $(x^2 - 4)(x^2 - 4)$
    d. $(x^2 - 2)(x^2 + 4)$

9. $(4x^2y^4)^{\frac{3}{2}}$ can be simplified to which of the following?

    a. $8x^3y^6$
    b. $4x^{\frac{5}{2}}y$
    c. $4xy$
    d. $32x^{\frac{7}{2}}y^{\frac{11}{2}}$

10. If $\sqrt{1 + x} = 4$, what is x?

    a. 10
    b. 15
    c. 20
    d. 25

11. Suppose $\frac{x+2}{x} = 2$. What is x?

    a. -1
    b. 0
    c. 2
    d. 4

12. A ball is thrown from the top of a high hill, so that the height of the ball as a function of time is $h(t) = -16t^2 + 4t + 6$, in feet. What is the maximum height of the ball in feet?

    a. 6
    b. 6.25
    c. 6.5
    d. 6.75

13. A rectangle has a length that is 5 feet longer than three times its width. If the perimeter is 90 feet, what is the length in feet?

    a. 10
    b. 20
    c. 25
    d. 35

14. Five students take a test. The scores of the first four students are 80, 85, 75, and 60. If the median score is 80, which of the following could NOT be the score of the fifth student?

    a. 60
    b. 80
    c. 85
    d. 100

15. In an office, there are 50 workers. A total of 60% of the workers are women, and the chances of a woman wearing a skirt is 50%. If no men wear skirts, how many workers are wearing skirts?

    a. 12
    b. 15
    c. 16
    d. 20

16. Ten students take a test. Five students get a 50. Four students get a 70. If the average score is 55, what was the last student's score?

    a. 20
    b. 40
    c. 50
    d. 60

17. A company invests \$50,000 in a building where they can produce saws. If the cost of producing one saw is \$40, then which function expresses the amount of money the company pays? The variable y is the money paid and x is the number of saws produced.

    a. $y = 50{,}000x + 40$
    b. $y + 40 = x - 50{,}000$
    c. $y = 40x - 50{,}000$
    d. $y = 40x + 50{,}000$

18. A six-sided die is rolled. What is the probability that the roll is 1 or 2?

    a. $\frac{1}{6}$
    b. $\frac{1}{4}$
    c. $\frac{1}{3}$
    d. $\frac{1}{2}$

19. A line passes through the origin and through the point (-3, 4). What is the slope of the line?

    a. $-\frac{4}{3}$
    b. $-\frac{3}{4}$
    c. $\frac{4}{3}$
    d. $\frac{3}{4}$

20. An equilateral triangle has a perimeter of 18 feet. If a square whose sides have the same length as one side of the triangle is built, what will be the area of the square?

    a. 6 square feet
    b. 36 square feet
    c. 256 square feet
    d. 1000 square feet

Reading

Directions for questions 1–9: Read the statement or passage and then choose the best answer to the question. Answer the question based on what is stated or implied in the statement or passage.

1. There are two major kinds of cameras on the market right now for amateur photographers. Camera enthusiasts can either purchase a digital single-lens reflex camera (DSLR) camera or a compact system camera (CSC). The main difference between a DSLR and a CSC is that the DSLR has a full-sized sensor, which means it fits in a much larger body. The CSC uses a mirrorless system, which makes for a lighter, smaller camera. While both take quality pictures, the DSLR generally has better picture quality due to the larger sensor. CSCs still take very good quality pictures and are more convenient to carry than a DSLR. This makes the CSC an ideal choice for the amateur photographer looking to step up from a point-and-shoot camera.

What is the main difference between the DSLR and CSC?
    a. The picture quality is better in the DSLR.
    b. The CSC is less expensive than the DSLR.
    c. The DSLR is a better choice for amateur photographers.
    d. The DSLR's larger sensor makes it a bigger camera than the CSC.

2. When selecting a career path, it's important to explore the various options available. Many students entering college may shy away from a major because they don't know much about it. For example, many students won't opt for a career as an actuary, because they aren't exactly sure what it entails. They would be missing out on a career that is very lucrative and in high demand. Actuaries work in the insurance field and assess risks and premiums. The average salary of an actuary is $100,000 per year. Another career option students may avoid, due to lack of knowledge of the field, is a hospitalist. This is a physician that specializes in the care of patients in a hospital, as opposed to those seen in private practices. The average salary of a hospitalist is upwards of $200,000. It pays to do some digging and find out more about these lesser-known career fields.

What is an *actuary*?
    a. A doctor who works in a hospital.
    b. The same as a hospitalist.
    c. An insurance agent who works in a hospital.
    d. A person who assesses insurance risks and premiums.

3. Hard water occurs when rainwater mixes with minerals from rock and soil. Hard water has a high mineral count, including calcium and magnesium. The mineral deposits from hard water can stain hard surfaces in bathrooms and kitchens as well as clog pipes. Hard water can stain dishes, ruin clothes, and

reduce the life of any appliances it touches, such as hot water heaters, washing machines, and humidifiers.

One solution is to install a water softener to reduce the mineral content of water, but this can be costly. Running vinegar through pipes and appliances and using vinegar to clean hard surfaces can also help with mineral deposits.

From this passage, what can be concluded?
   a. Hard water can cause a lot of problems for homeowners.
   b. Calcium is good for pipes and hard surfaces.
   c. Water softeners are easy to install.
   d. Vinegar is the only solution to hard water problems.

4. Coaches of kids' sports teams are increasingly concerned about the behavior of parents at games. Parents are screaming and cursing at coaches, officials, players, and other parents. Physical fights have even broken out at games. Parents need to be reminded that coaches are volunteers, giving up their time and energy to help kids develop in their chosen sport. The goal of kids' sports teams is to learn and develop skills, but it's also to have fun. When parents are out of control at games and practices, it takes the fun out of the sport.

From this passage, what can be concluded?
   a. Coaches are modeling good behavior for kids.
   b. Organized sports are not good for kids.
   c. Parents' behavior at their kids' games needs to change.
   d. Parents and coaches need to work together.

5. While scientists aren't entirely certain why tornadoes form, they have some clues into the process. Tornadoes are dangerous funnel clouds that occur during a large thunderstorm. When warm, humid air near the ground meets cold, dry air from above, a column of the warm air can be drawn up into the clouds. Winds at different altitudes blowing at different speeds make the column of air rotate. As the spinning column of air picks up speed, a funnel cloud is formed. This funnel cloud moves rapidly and haphazardly. Rain and hail inside the cloud cause it to touch down, creating a tornado. Tornadoes move in a rapid and unpredictable pattern, making them extremely destructive and dangerous. Scientists continue to study tornadoes to improve radar detection and warning times.

The main purpose of this passage is to do which of the following?
   a. Show why tornadoes are dangerous.
   b. Explain how a tornado forms.
   c. Compare thunderstorms to tornadoes.
   d. Explain what to do in the event of a tornado.

6. Many people are unsure of exactly how the digestive system works. Digestion begins in the mouth where teeth grind up food and saliva breaks it down, making it easier for the body to absorb. Next, the food moves to the esophagus, and it is pushed into the stomach. The stomach is where food is stored and broken down further by acids and digestive enzymes, preparing it for passage into the intestines. The small intestine is where the nutrients are taken from food and passed into the blood stream. Other essential organs like the liver, gall bladder, and pancreas aid the stomach in breaking down food and

absorbing nutrients. Finally, food waste is passed into the large intestine where it is eliminated by the body.

The purpose of this passage is to do which of the following?
a. Explain how the liver works.
b. Show why it is important to eat healthy foods.
c. Explain how the digestive system works.
d. Show how nutrients are absorbed by the small intestine.

7. Osteoporosis is a medical condition that occurs when the body loses bone or makes too little bone. This can lead to brittle, fragile bones that easily break. Bones are already porous, and when osteoporosis sets in, the spaces in bones become much larger, causing them to weaken. Both men and women can contract osteoporosis, though it is most common in women over age 50. Loss of bone can be silent and progressive, so it is important to be proactive in prevention of the disease.

The main purpose of this passage is to do which of the following?
a. Discuss some of the ways people contract osteoporosis.
b. Describe different treatment options for those with osteoporosis.
c. Explain how to prevent osteoporosis.
d. Define osteoporosis.

8. Vacationers looking for a perfect experience should opt out of Disney parks and try a trip on Disney Cruise Lines. While a park offers rides, characters, and show experiences, it also includes long lines, often very hot weather, and enormous crowds. A Disney Cruise, on the other hand, is a relaxing, luxurious vacation that includes many of the same experiences as the parks, minus the crowds and lines. The cruise has top-notch food, maid service, water slides, multiple pools, Broadway-quality shows, and daily character experiences for kids. There are also many activities, such as bingo, trivia contests, and dance parties that can entertain guests of all ages. The cruise even stops at Disney's private island for a beach barbecue with characters, waterslides, and water sports. Those looking for the Disney experience without the hassle should book a Disney cruise.

The main purpose of this passage is to do which of the following?
a. Explain how to book a Disney cruise.
b. Show what Disney parks have to offer.
c. Show why Disney parks are expensive.
d. Compare Disney parks to a Disney cruise.

9. As summer approaches, drowning incidents will increase. Drowning happens very quickly and silently. Most people assume that drowning is easy to spot, but a person who is drowning doesn't make noise or wave their arms. Instead, they will have their head back and their mouth open, with just the face out of the water. A person who is truly in danger of drowning is not able to wave their arms in the air or move much at all. Recognizing these signs of drowning can prevent tragedy.

The main purpose of this passage is to do which of the following?
   a. Explain the dangers of swimming.
   b. Show how to identify the signs of drowning.
   c. Explain how to be a lifeguard.
   d. Compare the signs of drowning.

The next question is based on the following conversation between a scientist and a politician.

> Scientist: Last year was the warmest ever recorded in the last 134 years. During that time period, the ten warmest years have all occurred since 2000. This correlates directly with the recent increases in carbon dioxide as large countries like China, India, and Brazil continue developing and industrializing. No longer do just a handful of countries burn massive amounts of carbon-based fossil fuels; it is quickly becoming the case throughout the whole world as technology and industry spread.

> Politician: Yes, but there is no causal link between increases in carbon emissions and increasing temperatures. The link is tenuous and nothing close to certain. We need to wait for all of the data before drawing hasty conclusions. For all we know, the temperature increase could be entirely natural. I believe the temperatures also rose dramatically during the dinosaurs' time, and I do not think they were burning any fossil fuels back then.

10. What is one point on which the scientist and politician agree?
   a. Burning fossil fuels causes global temperatures to rise.
   b. Global temperatures are increasing.
   c. Countries must revisit their energy policies before it's too late.
   d. Earth's climate naturally goes through warming and cooling periods.

The next question is based on the following passage.

> A famous children's author recently published a historical fiction novel under a pseudonym; however, it did not sell as many copies as her children's books. In her earlier years, she had majored in history and earned a graduate degree in Antebellum American History, which is the time frame of her new novel. Critics praised this newest work far more than the children's series that made her famous. In fact, her new novel was nominated for the prestigious Albert J. Beveridge Award but still isn't selling like her children's books, which fly off the shelves because of her name alone.

11. Which one of the following statements might be accurately inferred based on the above passage?
   a. The famous children's author produced an inferior book under her pseudonym.
   b. The famous children's author is the foremost expert on Antebellum America.
   c. The famous children's author did not receive the bump in publicity for her historical novel that it would have received if it were written under her given name.
   d. People generally prefer to read children's series than historical fiction.

The next three questions are based on the following passage.

Smoking is Terrible

Smoking tobacco products is terribly destructive. A single cigarette contains over 4,000 chemicals, including 43 known carcinogens and 400 deadly toxins. Some of the most dangerous

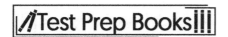

ingredients include tar, carbon monoxide, formaldehyde, ammonia, arsenic, and DDT. Smoking can cause numerous types of cancer including throat, mouth, nasal cavity, esophageal, gastric, pancreatic, renal, bladder, and cervical cancer.

Cigarettes contain a drug called nicotine, one of the most addictive substances known to man. Addiction is defined as a compulsion to seek the substance despite negative consequences. According to the National Institute of Drug Abuse, nearly 35 million smokers expressed a desire to quit smoking in 2015; however, more than 85 percent of those who struggle with addiction will not achieve their goal. Almost all smokers regret picking up that first cigarette. You would be wise to learn from their mistake if you have not yet started smoking.

According to the U.S. Department of Health and Human Services, 16 million people in the United States presently suffer from a smoking-related condition and nearly nine million suffer from a serious smoking-related illness. According to the Centers for Disease Control and Prevention (CDC), tobacco products cause nearly six million deaths per year. This number is projected to rise to over eight million deaths by 2030. Smokers, on average, die ten years earlier than their nonsmoking peers.

In the United States, local, state, and federal governments typically tax tobacco products, which leads to high prices. Nicotine users who struggle with addiction sometimes pay more for a pack of cigarettes than for a few gallons of gas. Additionally, smokers tend to stink. The smell of smoke is all-consuming and creates a pervasive nastiness. Smokers also risk staining their teeth and fingers with yellow residue from the tar.

Smoking is deadly, expensive, and socially unappealing. Clearly, smoking is not worth the risks.

12. Which of the following statements most accurately summarizes the passage?
    a. Tobacco is less healthy than many alternatives.
    b. Tobacco is deadly, expensive, and socially unappealing, and smokers would be much better off kicking the addiction.
    c. In the United States, local, state, and federal governments typically tax tobacco products, which leads to high prices.
    d. Tobacco products shorten smokers' lives by ten years and kill more than six million people per year.

13. The author would be most likely to agree with which of the following statements?
    a. Smokers should only quit cold turkey and avoid all nicotine cessation devices.
    b. Other substances are more addictive than tobacco.
    c. Smokers should quit for whatever reason that gets them to stop smoking.
    d. People who want to continue smoking should advocate for a reduction in tobacco product taxes.

14. Which of the following represents an opinion statement on the part of the author?
    a. According to the Centers for Disease Control and Prevention (CDC), tobacco products cause nearly six million deaths per year.
    b. Nicotine users who struggle with addiction sometimes pay more for a pack of cigarettes than a few gallons of gas.
    c. They also risk staining their teeth and fingers with yellow residue from the tar.
    d. Additionally, smokers tend to stink. The smell of smoke is all-consuming and creates a pervasive nastiness.

The next three questions are based on the following passage.

Christopher Columbus is often credited for discovering America. This is incorrect. First, it is impossible to "discover" something where people already live; however, Christopher Columbus did explore places in the New World that were previously untouched by Europe, so the term "explorer" would be more accurate. Another correction must be made, as well: Christopher Columbus was not the first European explorer to reach the present day Americas! Rather, it was Leif Erikson who first came to the New World and contacted the natives, nearly five hundred years before Christopher Columbus.

Leif Erikson, the son of Erik the Red (a famous Viking outlaw and explorer in his own right), was born in either 970 or 980, depending on which historian you seek. His own family, though, did not raise Leif, which was a Viking tradition. Instead, one of Erik's prisoners taught Leif reading and writing, languages, sailing, and weaponry. At age 12, Leif was considered a man and returned to his family. He killed a man during a dispute shortly after his return, and the council banished the Erikson clan to Greenland.

In 999, Leif left Greenland and traveled to Norway where he would serve as a guard to King Olaf Tryggvason. It was there that he became a convert to Christianity. Leif later tried to return home with the intention of taking supplies and spreading Christianity to Greenland, however his ship was blown off course and he arrived in a strange new land: present day Newfoundland, Canada.

When he finally returned to his adopted homeland Greenland, Leif consulted with a merchant who had also seen the shores of this previously unknown land we now know as Canada. The son of the legendary Viking explorer then gathered a crew of 35 men and set sail. Leif became the first European to touch foot in the New World as he explored present-day Baffin Island and Labrador, Canada. His crew called the land Vinland since it was plentiful with grapes.

During their time in present-day Newfoundland, Leif's expedition made contact with the natives whom they referred to as Skraelings (which translates to "wretched ones" in Norse). There are several secondhand accounts of their meetings. Some contemporaries described trade between the peoples. Other accounts describe clashes where the Skraelings defeated the Viking explorers with long spears, while still others claim the Vikings dominated the natives. Regardless of the circumstances, it seems that the Vikings made contact of some kind. This happened around 1000, nearly five hundred years before Columbus famously sailed the ocean blue.

Eventually, in 1003, Leif set sail for home and arrived at Greenland with a ship full of timber. In 1020, seventeen years later, the legendary Viking died. Many believe that Leif Erikson should receive more credit for his contributions in exploring the New World.

15. Which of the following is an opinion, rather than historical fact, expressed by the author?
    a. Leif Erikson was definitely the son of Erik the Red; however, historians debate the year of his birth.
    b. Leif Erikson's crew called the land Vinland since it was plentiful with grapes.
    c. Leif Erikson deserves more credit for his contributions in exploring the New World.
    d. Leif Erikson explored the Americas nearly five hundred years before Christopher Columbus.

16. Which of the following most accurately describes the author's main conclusion?
    a. Leif Erikson is a legendary Viking explorer.
    b. Leif Erikson deserves more credit for exploring America hundreds of years before Columbus.
    c. Spreading Christianity motivated Leif Erikson's expeditions more than any other factor.
    d. Leif Erikson contacted the natives nearly five hundred years before Columbus.

17. Which of the following can be logically inferred from the passage?
    a. The Vikings disliked exploring the New World.
    b. Leif Erikson's banishment from Iceland led to his exploration of present-day Canada.
    c. Leif Erikson never shared his stories of exploration with the King of Norway.
    d. Historians have difficulty definitively pinpointing events in the Vikings' history

This article discusses the famous poet and playwright William Shakespeare. Read it and answer questions 18-21.

People who argue that William Shakespeare is not responsible for the plays attributed to his name are known as anti-Stratfordians (from the name of Shakespeare's birthplace, Stratford-upon-Avon). The most common anti-Stratfordian claim is that William Shakespeare simply was not educated enough or from a high enough social class to have written plays overflowing with references to such a wide range of subjects like history, the classics, religion, and international culture. William Shakespeare was the son of a glove-maker, he only had a basic grade school education, and he never set foot outside of England—so how could he have produced plays of such sophistication and imagination? How could he have written in such detail about historical figures and events, or about different cultures and locations around Europe? According to anti-Stratfordians, the depth of knowledge contained in Shakespeare's plays suggests a well-traveled writer from a wealthy background with a university education, not a countryside writer like Shakespeare. But in fact, there is not much substance to such speculation, and most anti-Stratfordian arguments can be refuted with a little background about Shakespeare's time and upbringing.

First of all, those who doubt Shakespeare's authorship often point to his common birth and brief education as stumbling blocks to his writerly genius. Although it is true that Shakespeare did not come from a noble class, his father was a very *successful* glove-maker and his mother was from a very wealthy land owning family—so while Shakespeare may have had a country upbringing, he was certainly from a well-off family and would have been educated accordingly. Also, even though he did not attend university, grade school education in Shakespeare's time was actually quite rigorous and exposed students to classic drama through writers like Seneca and Ovid. It is not unreasonable to believe that Shakespeare received a very solid foundation in poetry and literature from his early schooling.

Next, anti-Stratfordians tend to question how Shakespeare could write so extensively about countries and cultures he had never visited before (for instance, several of his most famous works like *Romeo and Juliet* and *The Merchant of Venice* were set in Italy, on the opposite side of Europe!). But again, this criticism does not hold up under scrutiny. For one thing, Shakespeare was living in London, a bustling metropolis of international trade, the most populous city in England, and a political and cultural hub of Europe. In the daily crowds of people, Shakespeare would certainly have been able to meet travelers from other countries and hear firsthand accounts of life in their home country. And, in addition to the influx of information from world travelers, this was also the age of the printing press, a jump in technology that made it possible to print and circulate books much more easily than in the past. This also allowed for a freer flow of information across different countries, allowing people to read about life and ideas from throughout Europe. One needn't travel the continent in order to learn and write about its culture.

18. Which sentence contains the author's thesis?
    a. People who argue that William Shakespeare is not responsible for the plays attributed to his name are known as anti-Stratfordians.
    b. But in fact, there is not much substance to such speculation, and most anti-Stratfordian arguments can be refuted with a little background about Shakespeare's time and upbringing.
    c. It is not unreasonable to believe that Shakespeare received a very solid foundation in poetry and literature from his early schooling.
    d. Next, anti-Stratfordians tend to question how Shakespeare could write so extensively about countries and cultures he had never visited before.

19. In the first paragraph, "How could he have written in such detail about historical figures and events, or about different cultures and locations around Europe?" is an example of which of the following?
    a. Hyperbole
    b. Onomatopoeia
    c. Rhetorical question
    d. Appeal to authority

20. How does the author respond to the claim that Shakespeare was not well-educated because he did not attend university?
    a. By insisting upon Shakespeare's natural genius.
    b. By explaining grade school curriculum in Shakespeare's time.
    c. By comparing Shakespeare with other uneducated writers of his time.
    d. By pointing out that Shakespeare's wealthy parents probably paid for private tutors.

21. The word *bustling* in the third paragraph most nearly means which of the following?
    a. Busy
    b. Foreign
    c. Expensive
    d. Undeveloped

The next article is for questions 22-24.

The Myth of Head Heat Loss

It has recently been brought to my attention that most people believe that 75% of your body heat is lost through your head. I had certainly heard this before, and am not going to attempt to say I didn't believe it when I first heard it. It is natural to be gullible to anything said with enough authority. But the "fact" that the majority of your body heat is lost through your head is a lie.

Let me explain. Heat loss is proportional to surface area exposed. An elephant loses a great deal more heat than an anteater because it has a much greater surface area than an anteater. Each cell has mitochondria that produce energy in the form of heat, and it takes a lot more energy to run an elephant than an anteater.

So, each part of your body loses its proportional amount of heat in accordance with its surface area. The human torso probably loses the most heat, though the legs lose a significant amount as well. Some people have asked, "Why does it feel so much warmer when you cover your head than when you don't?" Well, that's because your head, because it is not clothed, is losing a lot of heat while the clothing on the rest of your body provides insulation. If you went outside with a hat and pants but no shirt, not only would you look silly, but your heat loss would be significantly greater because so much more of you would be exposed. So, if given the choice to cover your chest or your head in the cold, choose the chest. It could save your life.

22. Why does the author compare elephants and anteaters?
    a. To express an opinion.
    b. To give an example that helps clarify the main point.
    c. To show the differences between them.
    d. To persuade why one is better than the other.

23. Which of the following best describes the tone of the passage?
    a. Harsh
    b. Angry
    c. Casual
    d. Indifferent

24. The author appeals to which branch of rhetoric to prove their case?
    a. Factual evidence
    b. Emotion
    c. Ethics and morals
    d. Author qualification

Writing

Read the selection and answer questions 1-5.

[1]I have to admit that when my father bought an RV, I thought he was making a huge mistake. [2]In fact, I even thought he might have gone a little bit crazy. [3]I did not really know anything about recreational vehicles, but I knew that my dad was as big a "city slicker" as there was. [4]On trips to the beach, he preferred to swim at the pool, and whenever he went hiking, he avoided touching any plants for fear that they might be poison ivy. [5]Why would this man, with an almost irrational fear of the outdoors, want a 40-foot camping behemoth?

[6]The RV was a great purchase for our family and brought us all closer together. [7]Every morning we would wake up, eat breakfast, and broke camp. [8]We laughed at our own comical attempts to back The Beast into spaces that seemed impossibly small. [9]We rejoiced when we figured out how to "hack" a solution to a nagging technological problem. [10]When things inevitably went wrong and we couldn't solve the problems on our own, we discovered the incredible helpfulness and friendliness of the RV community. We even made some new friends in the process.

[11] Above all, owning the RV allowed us to share adventures travelling across America that we could not have experienced in cars and hotels. [12]Enjoying a campfire on a chilly summer evening with the mountains of Glacier National Park in the background, or waking up early in the

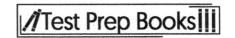

morning to see the sun rising over the distant spires of Arches National Park are memories that will always stay with me and our entire family. [13]Those are also memories that my siblings and I have now shared with our own children.

1. How should the author change sentence 11?
   a. Above all, it will allow us to share adventures travelling across America that we could not have experienced in cars and hotels.
   b. Above all, it allows you to share adventures travelling across America that you could not have experienced in cars and hotels.
   c. Above all, it allowed us to share adventures travelling across America that we could not have experienced in cars and hotels.
   d. Above all, it allows them to share adventures travelling across America that they could not have experienced in cars and hotels.

2. Which of the following examples would make a good addition to the selection after sentence 4?
   a. My father is also afraid of seeing insects.
   b. My father is surprisingly good at starting a campfire.
   c. My father negotiated contracts for a living.
   d. My father isn't even bothered by pigeons.

3. Which of the following would correct the error in sentence 7?
   a. Every morning we would wake up, ate breakfast, and broke camp.
   b. Every morning we would wake up, eat breakfast, and broke camp.
   c. Every morning we would wake up, eat breakfast, and break camp.
   d. Every morning we would wake up, ate breakfast, and break camp.

4. What transition word could be added to the beginning of sentence 6?
   a. Not surprisingly,
   b. Furthermore,
   c. As it turns out,
   d. Of course,

5. Which of the following topics would fit well between paragraph 1 and paragraph 2?
   a. A guide to RV holding tanks
   b. Describing how RV travel is actually not as outdoors-oriented as many think
   c. A description of different types of RVs
   d. Some examples of how other RV enthusiasts helped the narrator and his father during their travels

6. Which of the following is a clearer way to describe the following phrase?
   "employee-manager relations improvement guide"
   a. A guide to employing better managers
   b. A guide to improving relations between managers and employees
   c. A relationship between employees, managers, and improvement
   d. An improvement in employees' and managers' use of guides

Read the sentences, and then answer the following question.

7. Polls show that more and more people in the US distrust the government and view it as dysfunctional and corrupt. Every election, the same people are voted back into office.

Which word or words would best link these sentences?
    a. Not surprisingly,
    b. Understandably,
    c. And yet,
    d. Therefore,

8. Which of the following statements would make the best conclusion to an essay about civil rights activist Rosa Parks?
    a. On December 1, 1955, Rosa Parks refused to give up her bus seat to a white passenger, setting in motion the Montgomery bus boycott.
    b. Rosa Parks was a hero to many and came to symbolize the way that ordinary people could bring about real change in the Civil Rights Movement.
    c. Rosa Parks died in 2005 in Detroit, having moved from Montgomery shortly after the bus boycott.
    d. Rosa Parks' arrest was an early part of the Civil Rights Movement and helped lead to the passage of the Civil Rights Act of 1964.

Directions for questions 9-16: Select the best version of the underlined part of the sentence. The first choice is the same as the original sentence. If you think the original sentence is best, choose the first answer.

9. Since <u>none of the furniture were delivered</u> on time, we have to move in at a later date.
    a. none of the furniture were delivered
    b. none of the furniture was delivered
    c. all of the furniture were delivered
    d. all of the furniture was delivered

10. <u>An important issues stemming from this meeting</u> is that we won't have enough time to meet all of the objectives.
    a. An important issues stemming from this meeting
    b. Important issue stemming from this meeting
    c. An important issue stemming from this meeting
    d. Important issues stemming from this meeting

11. There were many questions <u>about what causes the case to have gone cold</u>, but the detective wasn't willing to discuss it with reporters.
    a. about what causes the case to have gone cold
    b. about why the case is cold
    c. about what causes the case to go cold
    d. about why the case went cold

12. The fact <u>the train set only includes four cars and one small track was a big disappointment</u> to my son.
   a. the train set only includes four cars and one small track was a big disappointment
   b. that the trains set only include four cars and one small track was a big disappointment
   c. that the train set only includes four cars and one small track was a big disappointment
   d. that the train set only includes four cars and one small track were a big disappointment

13. The rising popularity of the clean eating movement can be attributed <u>to the fact that experts say added sugars and chemicals in our food are to blame for the obesity epidemic.</u>
   a. to the fact that experts say added sugars and chemicals in our food are to blame for the obesity epidemic.
   b. in the facts that experts say added sugars and chemicals in our food are to blame for the obesity epidemic.
   c. to the fact that experts saying added sugars and chemicals in our food are to blame for the obesity epidemic.
   d. with the facts that experts say added sugars and chemicals in our food are to blame for the obesity epidemic.

14. She's looking for a suitcase that can fit all of her <u>clothes, shoes, accessory, and makeup.</u>
   a. clothes, shoes, accessory, and makeup.
   b. clothes, shoes, accessories, and makeup.
   c. clothes, shoes, accessories, and makeups.
   d. clothes, shoe, accessory, and makeup.

15. Shawn started taking guitar lessons <u>while he wanted to become a better musician.</u>
   a. while he wanted to become a better musician.
   b. because he wants to become a better musician.
   c. even though he wanted to become a better musician.
   d. because he wanted to become a better musician.

16. <u>Considering the recent rains we have had, it's a wonder</u> the plants haven't drowned.
   a. Considering the recent rains we have had, it's a wonder
   b. Consider the recent rains we have had, it's a wonder
   c. Considering for how much recent rain we have had, it's a wonder
   d. Considering, the recent rains we have had, it's a wonder

Directions for questions 17-20: Rewrite the sentence in your head following the directions given below. Keep in mind that your new sentence should be well written and should have essentially the same meaning as the original sentence.

17. There are many risks in firefighting, including smoke inhalation, exposure to hazardous materials, and oxygen deprivation, so firefighters are outfitted with many items that could save their lives, including a self-contained breathing apparatus.

Rewrite, beginning with <u>so firefighters</u>

The next words will be which of the following?
    a. are exposed to lots of dangerous situations.
    b. need to be very careful on the job.
    c. wear life-saving protective gear.
    d. have very risky jobs.

18. Though social media sites like Facebook, Instagram, and Snapchat have become increasingly popular, experts warn that teen users are exposing themselves to many dangers such as cyberbullying and predators.

Rewrite, beginning with <u>experts warn that</u>

The next words will be which of the following?
    a. Facebook is dangerous.
    b. they are growing in popularity.
    c. teens are using them too much.
    d. they can be dangerous for teens.

19. Student loan debt is at an all-time high, which is why many politicians are using this issue to gain the attention and votes of students, or anyone with student loan debt.

Rewrite, beginning with <u>Student loan debt is at an all-time high</u>

The next words will be which of the following?
    a. because politicians want students' votes.
    b. , so politicians are using the issue to gain votes.
    c. , so voters are choosing politicians who care about this issue.
    d. , and politicians want to do something about it.

20. Seasoned runners often advise new runners to get fitted for better quality running shoes because new runners often complain about minor injuries like sore knees or shin splints.

Rewrite, beginning with <u>Seasoned runners often advise new runners to get fitted for better quality running shoes</u>

The next words will be which of the following?

    a. to help them avoid minor injuries.
    b. because they know better.
    c. , so they can run further.
    d. to complain about running injuries.

Essay

Please read the prompt below and answer in an essay format in 300–600 words.

> Technology has been invading cars for the last several years, but there are some new high tech trends that are pretty amazing. It is now standard in many car models to have a rear-view camera, hands-free phone and text, and a touch screen digital display. Music can be streamed from a paired cell phone, and some displays can even be programmed with a personal photo. Sensors beep to indicate there is something in the driver's path when reversing and changing lanes. Rain-sensing windshield wipers and lights are automatic, leaving the driver with little to do but watch the road and enjoy the ride. The next wave of technology will include cars that automatically parallel park, and a self-driving car is on the horizon. These technological advances make it a good time to be a driver.

1. Analyze and evaluate the passage given.

2. State and develop your own perspective.

3. Explain the relationship between your perspective and the one given.

Answer Explanations for Practice Test #3

Math

1. B: First, subtract 4 from each side. This yields $6t = 12$. Now, divide both sides by 6 to obtain $t = 2$.

2. B: To be directly proportional means that $y = mx$. If x is changed from 5 to 20, the value of x is multiplied by 4. Applying the same rule to the y-value, also multiply the value of y by 4. Therefore, $y = 12$.

3. B: From the slope-intercept form, $y = mx + b$, it is known that b is the y-intercept, which is 1. Compute the slope as $\frac{2-1}{1-0} = 1$, so the equation should be $y = x + 1$.

4. A: Each bag contributes $4x + 1$ treats. The total treats will be in the form $4nx + n$ where n is the total number of bags. The total is in the form $60x + 15$, from which it is known $n = 15$.

5. D: Let a be the number of apples and b the number of bananas. Then, the total cost is $2a + 3b = 22$, while it also known that $a + b = 10$. Using the knowledge of systems of equations, cancel the b variables by multiplying the second equation by -3. This makes the equation $-3a - 3b = -30$. Adding this to the first equation, the b values cancel to get $-a = -8$, which simplifies to $a = 8$.

6. A: Finding the roots means finding the values of x when y is zero. The quadratic formula could be used, but in this case it is possible to factor by hand, since the numbers -1 and 2 add to 1 and multiply to -2. So, factor $x^2 + x - 2 = (x - 1)(x + 2) = 0$, then set each factor equal to zero. Solving for each value gives the values $x = 1$ and $x = -2$.

7. C: To find the y-intercept, substitute zero for x, which gives us $y = 0^{5/3} + (0 - 3)(0 + 1) = 0 + (-3)(1) = -3$.

8. A: This has the form $t^2 - y^2$, with $t = x^2$ and $y = 4$. It's also known that $t^2 - y^2 = (t + y)(t - y)$, and substituting the values for t and y into the right-hand side gives $(x^2 - 4)(x^2 + 4)$.

9. A: Simplify this to $(4x^2y^4)^{\frac{3}{2}} = 4^{\frac{3}{2}}(x^2)^{\frac{3}{2}}(y^4)^{\frac{3}{2}}$. Now, $4^{\frac{3}{2}} = (\sqrt{4})^3 = 2^3 = 8$. For the other, recall that the exponents must be multiplied, so this yields $8x^{2\times\frac{3}{2}}y^{4\times\frac{3}{2}} = 8x^3y^6$.

10. B: Start by squaring both sides to get $1 + x = 16$. Then subtract 1 from both sides to get $x = 15$.

11. C: Multiply both sides by x to get $x + 2 = 2x$, which simplifies to $-x = -2$, or $x = 2$.

12. B: The independent variable's coordinate at the vertex of a parabola (which is the highest point, when the coefficient of the squared independent variable is negative) is given by $x = -\frac{b}{2a}$. Substitute and solve for x to get $x = -\frac{4}{2(-16)} = \frac{1}{8}$. Using this value of x, the maximum height of the ball (y), can be calculated. Substituting x into the equation yields $h(t) = -16\frac{1}{8}^2 + 4\frac{1}{8} + 6 = 6.25$.

13. D: Denote the width as w and the length as l. Then, $l = 3w + 5$. The perimeter is $2w + 2l = 90$. Substituting the first expression for l into the second equation yields $2(3w + 5) + 2w = 90$, or $8w = 80$, so $w = 10$. Putting this into the first equation, it yields $l = 3(10) + 5 = 35$.

14. A: Lining up the given scores provides the following list: 60, 75, 80, 85, and one unknown. Because the median needs to be 80, it means 80 must be the middle data point out of these five. Therefore, the unknown data point must be the fourth or fifth data point, meaning it must be greater than or equal to 80. The only answer that fails to meet this condition is 60.

15. B: If 60% of 50 workers are women, then there are 30 women working in the office. If half of them are wearing skirts, then that means 15 women wear skirts. Since none of the men wear skirts, this means there are 15 people wearing skirts.

16. A: Let the unknown score be x. The average will be $\frac{5\times50+4\times70+x}{10} = \frac{530+x}{10} = 55$. Multiply both sides by 10 to get $530 + x = 550$, or $x = 20$.

17. D: For manufacturing costs, there is a linear relationship between the cost to the company and the number produced, with a y-intercept given by the base cost of acquiring the means of production, and a slope given by the cost to produce one unit. In this case, that base cost is $50,000, while the cost per unit is $40. So, $y = 40x + 50,000$.

18. C: A die has an equal chance for each outcome. Since it has six sides, each outcome has a probability of $\frac{1}{6}$. The chance of a 1 or a 2 is therefore $\frac{1}{6} + \frac{1}{6} = \frac{1}{3}$.

19. A: The slope is given by $m = \frac{y_2-y_1}{x_2-x_1} = \frac{0-4}{0-(-3)} = -\frac{4}{3}$.

20. B: An equilateral triangle has three sides of equal length, so if the total perimeter is 18 feet, each side must be 6 feet long. A square with sides of 6 feet will have an area of $6^2 = 36$ square feet.

Reading

1. D: The passage directly states that the larger sensor is the main difference between the two cameras. Choices A and B may be true, but these answers do not identify the major difference between the two cameras. Choice C states the opposite of what the paragraph suggests is the best option for amateur photographers, so it is incorrect.

2. D: An actuary assesses risks and sets insurance premiums. While an actuary does work in insurance, the passage does not suggest that actuaries have any affiliation with hospitalists or working in a hospital, so all other choices are incorrect.

3. A: The passage focuses mainly on the problems of hard water. Choice B is incorrect because calcium is not good for pipes and hard surfaces. The passage does not say anything about whether water softeners are easy to install, so Choice C is incorrect. Choice D is also incorrect because the passage does offer other solutions besides vinegar.

4. C: The main point of this paragraph is that parents need to change their poor behavior at their kids' sporting events. Choice A is incorrect because the coaches' behavior is not mentioned in the paragraph. Choice B suggests that sports are bad for kids, when the paragraph is about parents' behavior, so it is incorrect. While Choice D may be true, it offers a specific solution to the problem, which the paragraph does not discuss.

5. B: The main point of this passage is to show how a tornado forms. Choice A is off base because while the passage does mention that tornadoes are dangerous, it is not the main focus of the passage. While thunderstorms are mentioned, they are not compared to tornadoes, so Choice *C* is incorrect. Choice *D* is incorrect because the passage does not discuss what to do in the event of a tornado.

6. C: The purpose of this passage is to explain how the digestive system works. Choice *A* focuses only on the liver, which is a small part of the process and not the focus of the paragraph. Choice *B* is off-track because the passage does not mention healthy foods. Choice *D* only focuses on one part of the digestive system.

7. D: The main point of this passage is to define osteoporosis. Choice *A* is incorrect because the passage does not list ways that people contract osteoporosis. Choice *B* is incorrect because the passage does not mention any treatment options. While the passage does briefly mention prevention, it does not explain how, so Choice *C* is incorrect.

8. D: The passage compares Disney cruises with Disney parks. It does not discuss how to book a cruise, so Choice *A* is incorrect. Choice *B* is incorrect because though the passage does mention some of the park attractions, it is not the main point. The passage does not mention the cost of either option, so Choice *C* is incorrect.

9. B: The point of this passage is to show what drowning looks like. Choice *A* is incorrect because while drowning is a danger of swimming, the passage doesn't include any other dangers. The passage is not intended for lifeguards specifically, but for a general audience, so Choice *C* is incorrect. There are a few signs of drowning, but the passage does not compare them; thus, Choice *D* is incorrect.

10. B: The scientist and politician largely disagree, but the question asks for a point where the two are in agreement. The politician would not concur that burning fossil fuels causes global temperatures to rise; thus, Choice *A* is wrong. The politician also would not agree with Choice *C* suggesting that countries must revisit their energy policies. By inference from the given information, the scientist would likely not concur that earth's climate naturally goes through warming and cooling cycles; so Choice *D* is incorrect. However, both the scientist and politician would agree that global temperatures are increasing. The reason for this is in dispute. The politician thinks it is part of the earth's natural cycle; the scientist thinks it is from the burning of fossil fuels. However, both acknowledge an increase, so Choice *B* is the correct answer.

11. C: We are looking for an inference—a conclusion that is reached on the basis of evidence and reasoning—from the passage that will likely explain why the famous children's author did not achieve her usual success with the new genre (despite the book's acclaim). Choice *A* is wrong because the statement is false according to the passage. Choice *B* is wrong because, although the passage says the author has a graduate degree on the subject, it would be an unrealistic leap to infer that she is the foremost expert on Antebellum America. Choice *D* is wrong because there is nothing in the passage to lead us to infer that people generally prefer a children's series to historical fiction. In contrast, Choice *C* can be logically inferred since the passage speaks of the great success of the children's series and the declaration that the fame of the author's name causes the children's books to "fly off the shelves." Thus, she did not receive any bump from her name since she published the historical novel under a pseudonym, and Choice *C* is correct.

12. B: The author is opposed to tobacco. He cites disease and deaths associated with smoking. He points to the monetary expense and aesthetic costs. Choice *A* is incorrect because alternatives to smoking are not even addressed in the passage. Choice *C* is incorrect because it does not summarize the passage but rather is just a premise. Choice *D* is incorrect because, while these statistics are a premise in the argument, they do not represent a summary of the piece. Choice *B* is the correct answer because it states the three critiques offered against tobacco and expresses the author's conclusion.

13. C: We are looking for something the author would agree with, so it should be anti-smoking or an argument in favor of quitting smoking. Choice *A* is incorrect because the author does not speak against means of cessation. Choice *B* is incorrect because the author does not reference other substances but does speak of how addictive nicotine, a drug in tobacco, is. Choice *D* is incorrect because the author would not encourage reducing taxes to encourage a reduction of smoking costs, thereby helping smokers to continue the habit. Choice *C* is correct because the author is attempting to persuade smokers to quit smoking.

14. D: Here, we are looking for an opinion of the author's rather than a fact or statistic. Choice *A* is incorrect because quoting statistics from the Centers of Disease Control and Prevention is stating facts, not opinions. Choice *B* is incorrect because it expresses the fact that cigarettes sometimes cost more than a few gallons of gas. It would be an opinion if the author said that cigarettes were not affordable. Choice *C* is incorrect because yellow stains are a known possible adverse effect of smoking. Choice *D* is correct as an opinion because smell is subjective. Some people might like the smell of smoke, they might not have working olfactory senses, and/or some people might not find the smell of smoke akin to "pervasive nastiness," so this is the expression of an opinion. Thus, Choice *D* is the correct answer.

15. C: Choice *A* is incorrect because it describes facts: Leif Erikson was the son of Erik the Red and historians debate Leif's date of birth. These are not opinions. Choice *B* is incorrect; that Erikson called the land Vinland is a verifiable fact as is Choice *D* because he did contact the natives almost 500 years before Columbus. Choice *C* is the correct answer because it is the author's opinion that Erikson deserves more credit. That, in fact, is his conclusion in the piece, but another person could argue that Columbus or another explorer deserves more credit for opening up the New World to exploration. Rather than being an incontrovertible fact, it is a subjective value claim.

16. B: Choice *A* is incorrect because the author aims to go beyond describing Erikson as a mere legendary Viking. Choice *C* is incorrect because the author does not focus on Erikson's motivations, let alone name the spreading of Christianity as his primary objective. Choice *D* is incorrect because it is a premise that Erikson contacted the natives 500 years before Columbus, which is simply a part of supporting the author's conclusion. Choice *B* is correct because, as stated in the previous answer, it accurately identifies the author's statement that Erikson deserves more credit than he has received for being the first European to explore the New World.

17. D: Choice *A* is incorrect because the author never addresses the Vikings' state of mind or emotions. Choice *B* is incorrect because the author does not elaborate on Erikson's exile and whether he would have become an explorer if not for his banishment. Choice *C* is incorrect because there is not enough information to support this premise. It is unclear whether Erikson informed the King of Norway of his finding. Although it is true that the King did not send a follow-up expedition, he could have simply chosen not to expend the resources after receiving Erikson's news. It is not possible to logically infer whether Erikson told him. Choice *D* is correct because there are two examples—Leif Erikson's date of birth and what happened during the encounter with the natives—of historians having trouble pinning down important details in Viking history.

18. B: But in fact, there is not much substance to such speculation, and most anti-Stratfordian arguments can be refuted with a little background about Shakespeare's time and upbringing. The thesis is a statement that contains the author's topic and main idea. The main purpose of this article is to use historical evidence to provide counterarguments to anti-Stratfordians. Choice *A* is simply a definition; Choice *C* is a supporting detail, not a main idea; and Choice *D* represents an idea of anti-Stratfordians, not the author's opinion.

19. C: Rhetorical question. This requires readers to be familiar with different types of rhetorical devices. A rhetorical question is a question that is asked not to obtain an answer but to encourage readers to more deeply consider an issue.

20. B: By explaining grade school curriculum in Shakespeare's time. This question asks readers to refer to the organizational structure of the article and demonstrate understanding of how the author provides details to support their argument. This particular detail can be found in the second paragraph: "even though he did not attend university, grade school education in Shakespeare's time was actually quite rigorous."

21. A: Busy. This is a vocabulary question that can be answered using context clues. Other sentences in the paragraph describe London as "the most populous city in England" filled with "crowds of people," giving an image of a busy city full of people. Choice *B* is incorrect because London was in Shakespeare's home country, not a foreign one. Choice *C* is not mentioned in the passage. Choice *D* is not a good answer choice because the passage describes how London was a popular and important city, probably not an undeveloped one.

22. B: Choice *B* is correct because the author is trying to demonstrate the main idea, which is that heat loss is proportional to surface area, and so they compare two animals with different surface areas to clarify the main point. Choice *A* is incorrect because the author uses elephants and anteaters to prove a point, that heat loss is proportional to surface area, not to express an opinion. Choice *C* is incorrect because though the author does use them to show differences, they do so in order to give examples that prove the above points, so Choice *C* is not the best answer. Choice *D* is incorrect because there is no language to indicate favoritism between the two animals.

23. C: Because of the way that the author addresses the reader, and also the colloquial language that the author uses (i.e., "let me explain," "so," "well," didn't," "you would look silly," etc.), Choice *C* is the best answer because it has a much more casual tone than the usual informative article. Choice *A* may be a tempting choice because the author says the "fact" that most of one's heat is lost through their head is a "lie," and that someone who does not wear a shirt in the cold looks silly, but it only happens twice within all the diction of the passage and it does not give an overall tone of harshness. Choice *B* is incorrect because again, while not necessarily nice, the language does not carry an angry charge. The author is clearly not indifferent to the subject because of the passionate language that they use, so Choice *D* is incorrect.

24. A: The author gives logical examples and reasons in order to prove that most of one's heat is not lost through their head, therefore Choice *A* is correct. Choice *B* is incorrect because there is not much emotionally charged language in this selection, and even the small amount present is greatly outnumbered by the facts and evidence. Choice *C* is incorrect because there is no mention of ethics or morals in this selection. Choice *D* is incorrect because the author never qualifies himself as someone who has the authority to be writing on this topic.

Writing

1. C: The sentence should be in the same tense and person as the rest of the selection. The rest of the selection is in past tense and first person. Choice *A* is in future tense. Choice *B* is in second person. Choice *D* is in third person. While none of these sentences are incorrect by themselves, they are written in a tense that is different from the rest of the selection. Only Choice *C* maintains tense and voice consistent with the rest of the selection.

2. A: Choices *B* and *D* go against the point the author is trying to make—that the father is not comfortable in nature. Choice *C* is irrelevant to the topic. Choice *A* is the only choice that emphasizes the father's discomfort with spending time in nature.

3. C: This sentence uses verbs in a parallel series, so each verb must follow the same pattern. In order to fit with the helping verb "would," each verb must be in the present tense. In Choices *A*, *B*, and *D*, one or more of the verbs switches to past tense. Only Choice *C* remains in the same tense, maintaining the pattern.

4. C: In paragraph 2, the author surprises the reader by asserting that the opposite of what was expected was in fact true—the city slicker father actually enjoyed the RV experience. Only Choice *C* indicates this shift in expected outcome, while the other choices indicate a continuation of the previous expectation.

5. B: Choices *A* and *C* are irrelevant to the topic. They deal more with details about RVs while the author is more concerned with the family's experiences with them. Choice *D* is relevant to the topic, but it would fit better between paragraphs 2 and 3, since the author does not mention this point until the end of the second paragraph. Choice *B* would help explain to the reader why the father, who does not enjoy the outdoors, could end up enjoying RVs so much.

6. B: Stacked modifying nouns such as this example are untangled by starting from the end and adding words as necessary to provide meaning. In this case, a *guide* to *improving relations* between *managers* and *employees*. Choices *C* and *D* do not define the item first as a guide. Choice *A* does identify as a guide, but confuses the order of the remaining descriptors. Choice *B* is correct, as it unstacks the nouns in the correct order and also makes logical sense.

7. C: The second sentence tells of an unexpected outcome of the first sentence. Choice *A*, Choice *B*, and Choice *D* indicate a logical progression, which does not match this surprise. Only Choice *C* indicates this unexpected twist.

8. B: Choice *A*, Choice *C*, and Choice *D* all relate facts but do not present the kind of general statement that would serve as an effective summary or conclusion. Choice *B* is correct.

9. B: Choice *A* uses the plural form of the verb, when the subject is the pronoun *none*, which needs a singular verb. Choice *C* also uses the wrong verb form and uses the word *all* in place of *none*, which doesn't make sense in the context of the sentence. Choice *D* uses *all* again, and is missing the comma, which is necessary to set the dependent clause off from the independent clause.

10. C: In this answer, the article and subject agree, and the subject and predicate agree. Choice *A* is incorrect because the article *an* and *issues* do not agree in number. Choice *B* is incorrect because an article is needed before *important issue*. Choice *D* is incorrect because the plural subject *issues* does not agree with the singular verb *is*.

11. D: Choices *A* and *C* use additional words and phrases that aren't necessary. Choice *B* is more concise but uses the present tense of *is*. This does not agree with the rest of the sentence, which uses past tense. The best choice is Choice *D*, which uses the most concise sentence structure and is grammatically correct.

12. C: Choice *A* is missing the word *that*, which is necessary for the sentence to make sense. Choice *B* pluralizes *trains* and uses the singular form of the word *include*, so it does not agree with the word *set*. Choice *D* changes the verb to *were*, which is in plural form and does not agree with the singular subject.

13. A: Choices *B* and *D* both use the expression *attributed to the fact* incorrectly. It can only be attributed *to* the fact, not *with* or *in* the fact. Choice *C* incorrectly uses a gerund, *saying,* when it should use the present tense of the verb *say*.

14. B: Choice *B* is correct because it uses correct parallel structure of plural nouns. Choice *A* is incorrect because the word *accessory* is in singular form. Choice *C* is incorrect because it pluralizes *makeup*, which is already in plural form. Choice *D* is incorrect because it again uses the singular *accessory*, and it uses the singular *shoe*.

15. D: In a cause/effect relationship, it is correct to use the word because in the clausal part of the sentence. This can eliminate both Choices *A* and *C* which don't clearly show the cause/effect relationship. Choice *B* is incorrect because it uses the present tense, when the first part of the sentence is in the past tense. It makes grammatical sense for both parts of the sentence to be in present tense.

16. A: In Choice *B*, the present tense form of the verb *consider* creates an independent clause joined to another independent clause with only a comma, which is a comma splice and grammatically incorrect. Choices *C* and *D* use the possessive form of *its*, when it should be the contraction *it's* for *it is*. Choice *D* also includes incorrect comma placement.

17. C: The original sentence states that firefighting is dangerous, making it necessary for firefighters to wear protective gear. The portion of the sentence that needs to be rewritten focuses on the gear, not the dangers, of firefighting. Choices *A*, *B*, and *D* all discuss the danger, not the gear, so *C* is the correct answer.

18. D: The original sentence states that though the sites are popular, they can be dangerous for teens, so *D* is the best choice. Choice *A* does state that there is danger, but it doesn't identify teens and limits it to just one site. Choice *B* repeats the statement from the beginning of the sentence, and Choice *C* says the sites are used too much, which is not the point made in the original sentence.

19. B: The original sentence focuses on how politicians are using the student debt issue to their advantage, so Choice *B* is the best answer choice. Choice *A* says politicians want students' votes but suggests that it is the reason for student loan debt, which is incorrect. Choice *C* shifts the focus to voters, when the sentence is really about politicians. Choice *D* is vague and doesn't best restate the original meaning of the sentence.

20. A: This answer best matches the meaning of the original sentence, which states that seasoned runners offer advice to new runners because they have complaints of injuries. Choice *B* may be true, but it doesn't mention the complaints of injuries by new runners. Choice *C* may also be true, but it does not match the original meaning of the sentence. Choice *D* does not make sense in the context of the sentence.

Dear TSI Test Taker,

We would like to start by thanking you for purchasing this study guide for your TSI exam. We hope that we exceeded your expectations.

Our goal in creating this study guide was to cover all of the topics that you will see on the test. We also strove to make our practice questions as similar as possible to what you will encounter on test day. With that being said, if you found something that you feel was not up to your standards, please send us an email and let us know.

We would also like to let you know about other books in our catalog that may interest you.

ACCUPLACER

This can be found on Amazon: amazon.com/dp/1628459344

SAT

amazon.com/dp/1628457376

ACT

amazon.com/dp/1628454709

AP Biology

amazon.com/dp/1628454989

SAT Math 1

amazon.com/dp/1628459891

We have study guides in a wide variety of fields. If the one you are looking for isn't listed above, then try searching for it on Amazon or send us an email.

Thanks Again and Happy Testing!
Product Development Team
info@studyguideteam.com

Interested in buying more than 10 copies of our product? Contact us about bulk discounts:

bulkorders@studyguideteam.com

FREE Test Taking Tips DVD Offer

To help us better serve you, we have developed a Test Taking Tips DVD that we would like to give you for FREE. **This DVD covers world-class test taking tips that you can use to be even more successful when you are taking your test.**

All that we ask is that you email us your feedback about your study guide. Please let us know what you thought about it – whether that is good, bad or indifferent.

To get your **FREE Test Taking Tips DVD**, email freedvd@studyguideteam.com with "FREE DVD" in the subject line and the following information in the body of the email:

   a. The title of your study guide.

   b. Your product rating on a scale of 1-5, with 5 being the highest rating.

   c. Your feedback about the study guide. What did you think of it?

   d. Your full name and shipping address to send your free DVD.

If you have any questions or concerns, please don't hesitate to contact us at freedvd@studyguideteam.com.

Thanks again!